JOHN MILTON

Samson Agonistes, Sonnets, &c.

THE CAMBRIDGE MILTON FOR
SCHOOLS AND COLLEGES

GENERAL EDITOR: J. B. BROADBENT

Already published:

John Milton: introductions, edited by John Broadbent

Odes, Pastorals, Masques, edited by
David Aers, John Broadbent, Winifred Maynard,
Peter Mendes, and Lorna Sage

Paradise Lost: introduction, John Broadbent

Paradise Lost: books I–II, edited by John Broadbent

Paradise Lost: books III–IV, edited by Lois Potter and John Broadbent

Paradise Lost: books V–VI, edited by
Robert Hodge and Isabel MacCaffrey

Paradise Lost: books VII–VIII, edited by
David Aers and Mary Ann Radzinowicz

Paradise Lost: books IX–X, edited by J. Martin Evans

Paradise Lost: books XI–XII, edited by
Michael Hollington with Lawrence Wilkinson

JOHN MILTON
Samson Agonistes, Sonnets, &c.

Edited by
JOHN BROADBENT and ROBERT HODGE
University of East Anglia, Norwich

with translations of selected Latin and Italian poetry by
ROBERT HODGE

I molti, immani, sparsi, grigi sassi
Frementi ancora alle segrete fionde
Di originarie fiamme soffocate
Od ai terrori di fiumane vergini
Ruinanti in implacabili carezze;
—Sopra l'abbaglio della sabbia rigidi
In un vuoto orizzonte, non rammenti?
E la recline, che s'apriva all'unico
Raccogliersi dell'ombra nella valle,
Araucaria, anelando ingigantita,
Volta nell'ardua selce d'erme fibre
Più delle altre dannate refrattaria,
Fresca la bocca di farfalle e d'erbe
Dove dalle radici si tagliava

. . . . . .

Grazia felice,
Non avresti potuto non spezzarti
In una cecità tanto indurita
Tu semplice soffio e cristallo,

Troppo umano lampo per l'empio,
Selvoso, accanito, ronzante
Ruggito d'un sole ignudo.

GUISEPPE UNGARETTI, from *Tu ti spezzasti*
in his *Il dolore* Arnoldo Mondadori Editore 1947

Cambridge University Press

Cambridge
London · New York · Melbourne

CAMBRIDGE UNIVERSITY PRESS
Cambridge, New York, Melbourne, Madrid, Cape Town, Singapore, São Paulo, Delhi

Cambridge University Press
The Edinburgh Building, Cambridge CB2 8RU, UK

Published in the United States of America by Cambridge University Press, New York

www.cambridge.org
Information on this title: www.cambridge.org/9780521214742

First published 1977
Re-issued in this digitally printed version 2009

A catalogue record for this publication is available from the British Library

Library of Congress Cataloguing in Publication data
Milton, John, 1608–1674.
John Milton : Samson Agonistes, sonnets, etc.
(The Cambridge Milton for schools and colleges)
I. Broadbent, John Barclay. II. Hodge, Robert Ian Vere. III. Series.
PR3552.B76   821'.4   76–28003

ISBN 978-0-521-05734-9 hardback
ISBN 978-0-521-21474-2 paperback

Acknowledgements

The photograph on the cover shows a detail from *Samson and Delilah* by Andrea Mantegna *c.* 1495, and is reproduced by courtesy of the Trustees, The National Gallery, London, where the painting may be seen. It is about eighteen inches (457 mm) high and imitates the appearance of a bas-relief laid on a ground of coloured marble. Cut into the branch of the tree is the medieval proverb *Foemina Diabolo tribus assibus est mala peior* – a bad woman is three times worse than the devil.

The extract on the title-page, reproduced by permission of Arnoldo Mondadori Editore, refers to the death of the poet's son when Ungaretti was in Brazil as Professor of Italian literature at São Paulo 1936–42; perhaps the South American landscape also refers to the violence of the world of men. It may be translated:

> All those huge dispersed grey rocks
> Shake still from the weird ballistics
> Of the primordial fires, now damped,
> Or terrors of the virgin floods
> Ruining in irresistible lust –
> Don't you remember them, stark on the dazzle of sand
> Against an empty horizon?
> And leaning, spread in the shade
> Which it alone gathered into that valley,
> The Araucaria pine, gigantically yearning,
> Its separate fibres fused into ardent flint,
> Refractory above all plants, resistant and damned –
> Its mouth, where the trunk split away from the root,
> Delicious with butterflies and grass
> .   .   .   .

O happy grace,
That could not have not been broken
In so stony a blindness –
The pure you of breath and crystal:

O blaze too human for sin,
Brutal bristled slavering
Roar of the naked sun.

The genus *Araucaria* includes the monkey-puzzle tree. For an introduction to Ungaretti and this poem see Stanley Burnshaw ed. *The poem itself: 150 European poems translated and analysed* 1960, Pelican 1964.

Contents

TEXTS OF THE SHORTER POEMS, WITH

TRANSLATIONS AND FOOTNOTES

Samson agonistes

The editors

The shorter poems edited, and the Latin and Italian translated, by

ROBERT HODGE Western Australia (where he studied classics and English, trained as a teacher and taught for a while), Cambridge (fellow of Churchill College), East Anglia. Has conducted research on poetry and revolutions in the 17th century: Latin elegy, especially Propertius: now working on language and ideology, and literature and politics. Dr Hodge is married to a potter, and is now senior lecturer at the new Australian University, Murdoch.

Samson Agonistes edited by

JOHN BROADBENT Edinburgh, Cambridge (fellow of King's College), East Anglia. Author of *Some graver subject: an essay on PL*; *Milton: Comus and Samson*; *Poetic love*. General editor of this series and editor of other volumes in it. Ed. Smart's *Song to David*, and 2-volume anthology *Poets of the 17th century*. Programmes on art history and mythology for BBC TV in the 1960s. Professor Broadbent is married to a social worker.

Foreword

This volume is part of the Cambridge Milton s eries. It can be used independently but we assume that you refer as appropriate to two other volumes in particular:

(1) *John Milton: introductions*: a collaborative volume ed. J. B. Broadbent. For Milton's life, times, ideas; music, visual arts, science, use of the Bible; place in literary history. That volume also contains a *General introduction to the early poems* by Lorna Sage which is intended to serve both for the poems published here, and those in

(2) Milton's *Odes, pastorals, masques*: a collaborative volume containing *Nativity ode, Passion, Circumcision, Time, Solemn music, L'allegro* and *Il penseroso, Arcades* and *Comus, Lycidas,* with full introductory material and a chronology. Some of the poems in that volume are contemporaneous with some in this; and there are important links, eg between *Lycidas* in that volume and *Epitaphium Damonis* in this, as pastoral elegies; between *Comus* there and *Samson* here as early and late Miltonic dramas.

In this volume our first principle of selection has been to include virtually all Milton's shorter English poems not printed in *Odes, pastorals, masques*; we have also included some Latin and nearly all his Italian poems. However the Cambridge Milton series does not purport to be complete; in particular we regret that economy forces us to omit some Latin poetry which is superior to some of the English juvenilia; and allows us to include only one of Milton's translations.

The series will supersede A. W. Verity's Pitt Press edition of Milton's poetry published from Cambridge 1891 *et seq*. It is designed for use by the individual student, and the class, and the teacher, in schools and colleges, from about the beginning of the sixth form to the end of the first postgraduate year course in England. Introductions and notes aim to provide enough material for the reader to work on for himself, but nothing of a professionally academic kind. We hope that if any volume of text is prescribed for examination, some of its contents will not be set, but left for the student to explore at will.

We have also considered him as a poet, and such he was, if ever human nature could boast it had produced one...in expounding him we have therefore always given, as well as we were enabled, a poetic sense...for a poem, such a one as this especially, is not to be read, and construed, as an Act of Parliament, or a mathematical dissertation; the things of the spirit are spiritually discerned.

JONATHAN RICHARDSON father and son Explanatory notes
and remarks on Paradise Lost 1734

For further suggestions about using the series, and reading the texts, see the Preface to the Cambridge Milton printed in *Paradise lost* volumes of the series.

The texts are based on the latest editions published in Milton's lifetime: ie *Poems of Mr John Milton, both English and Latin* 1645 and its enlarged second edition *Poems, &c. upon Several Occasions* 1673 which is the chief source and the only one for some of the shorter poems; and the 1671 edition of *Samson agonistes*, which was published together with *Paradise regained*. But the text as such has no authority. The translations have been prepared specifically for this volume.

Spelling has been modernized (except where it would completely alter pronunciation, eg *anow* has been changed to *enow* but not to *enough*).

Stress marks (′) have been added where Milton seems to have intended a stress unusual for us. Grave accents (`) have been added to indicate voiced syllables in such cases as *blessèd* and in unfamiliar names.

Milton showed much elision of *e*'s, eg 'th'heavens'. These have been omitted because the elision comes more naturally if we read it with our usual neutral *e* sounds in such cases, than if we try to say *theavens*.

On the other hand, Milton's punctuation has been left almost untouched. It is not the same as ours, but you soon get used to it, and to tamper would alter the rhythm.

Biographical references. The place of publication is not cited for works published in London or New York.

xiv

Introduction to the shorter poems

How to read peripheral poetry

The poetry in this part of the volume can seem very unsatisfactory if read in the wrong kind of way. You can't easily focus on the poems for their own sakes as self-contained works of art, or use them as the starting point for excursions which continually return to the poem as a fixed point (as you could with *Samson*). These poems are interesting precisely because they resist such an approach. They're too peripheral, ancillary to some other area of interest, existing on some boundary. Much of it is early verse, existing on the boundary between youth and maturity, full of false starts, with themes and concerns, problems and commitments of the mature poet already tentatively emerging. The interest shifts from the poems themselves to the creative process, following the sometimes erratic course of the poet's development, seeing him when he's open, vulnerable, not quite sure where he's going next. It becomes interesting then to know where he did in fact go next. This requires a kind of bifocal reading, a movement between early work and mature achievement, promise and fulfilment. It's important to be surprised as well as to have your expectations confirmed. Looking at *Samson*, for instance, would you have predicted M's early erotic verse? Partly because M was consciously in the process of discovering himself as a poet, these poems also contain most of M's poetic reflection on the role of the poet and the nature of his own poetry.

Much of it is occasional poetry, poetry written on particular occasions. Such poetry exists on another boundary, between literature and life. Its meaning often depends on knowing the situation it grew out of: who it was about or addressed to, what

M was doing or thinking at the time, what was the state of the nation. Some history becomes indispensable.

With the Latin and Italian poetry even more difficulties arise. The boundary they lie on is one of language, M's explorations of other languages and other cultures. If your Latin or Italian isn't strong, you'll have to rely on translations, notes and a dictionary, moving between English and the other language; remembering that M was not a native speaker of Latin or Italian, either.

So reading this kind of poetry, you can't make neat distinctions between poetry and background. From some points of view, it's all background. Since it comes in the same volume as *Samson* it could be read as a long appendix to that major poem. Hence there's no separate appendix to this section: that would be like having an appendix attached to an appendix. Linking references throughout are designed to help the juxtaposition. But the poems shouldn't be allowed to dissolve into a surrounding penumbra. To see the connections between literature and life which are more transparent and urgent in these poems requires a fully alert and totally engaged reading. The M who emerges is not less interesting for being more comprehensible, accessible and human. Finally even a *Samson* or *PL* can gain from being read as peripheral.

Further reading
JM: introductions 1973 in this series is especially relevant for guidance with background: particularly its chronology, Broadbent *Inside M*, Rivers *M's life and times* and Sage *M's early poems: a general introduction*. Diekhoff *M on himself* 1939 collects conveniently together autobiographical fragments from M's writings. The two major biographies, Masson's 8-volume Victorian life (*Life of JM* 1858–80) and Parker's substantial recent biography (*Milton* Oxford 1968) should not be avoided for their bulk and detail. Use their index and be clear what you are looking for (have specific questions about how specific details of poems grow out their context) and they can be valuable aids.

Dates

Dates are clearly important with such poetry, because they help to locate the poem in the experience it grew out of, which it most immediately comments on. They also give an objective ordering of the poems. If two poems touch on the same theme, the second was written by a poet who had written the first, and is now either confirming or contradicting what he once said. There is inevitably a dialectical relation between the two, whether or not the poet is aware of it.

But dates are not always available, or reliable. You can't even trust M's statements about them. He put dates or his age on many of his early poems, but demonstrably got them wrong so often that we might suspect him of habitually lying about his age. After all, they were offered as juvenile verse, and he may have wished to seem a child prodigy. Or perhaps he wished to dissociate himself from some of the attitudes they expressed, particularly the eroticism and royalist or establishment sentiments of some of the early verse. Coleridge is another poet who was often misleading about the real course of his poetic development.

In many cases, the date of a poem can only be guessed at. Here the scholar may have something to contribute but cannot be decisive. This uncertainty hands over the immensely creative task of dating to the student. The grounds of the decision are your sense of work itself, and everything you know about the relevant history. The result is not merely a date, but a new grasp of the possibilities of the bio-aesthetic whole. Take *Sonnet 16*, on his blindness, for instance. The dates commonly suggested are 1648–50, or post-1660. What biographical material feels most relevant? At the earlier date, M was once-married, a propagandist for a fragile but still successful revolution, in process of losing his sight. At the later date he was totally blind, perhaps married to his third wife (from 1663), living with the total failure of his revolution. Or perhaps parts of the poem were written at both times, or written earlier and revised later? Is it or does it feel the same poem in each of these cases?

A story by Jorge Luis Borges suggests that it isn't. In *Pierre Menard, author of 'Don Quixote'* Borges envisages a French symbolist poet and philosopher who sets himself the task of rewriting Cervantes' famous novel using Cervantes' exact words. The result, Borges' narrator claims, is a wholly new work in its value and effect. What was natural for the 17th-century Spaniard to write and believe is inevitably anachronistic, ironic and contrived for the sophisticated 20th-century Frenchman, who has had to renounce so much of his own history and culture in order to write and endorse the exact words of the earlier work. Borges ends the story with a general comment:

Menard (perhaps without wishing to) has enriched, by means of a new technique, the hesitant and rudimentary art of reading; the technique is one of deliberate anachronism and erroneous attributions. This technique, with its infinite applications, urges us to run through the *Odyssey* as if it were written after the *Aeneid*, and to read *Le Jardin du Centaure* by Madame Henri Bachelier as though it were by Madame Henri Bachelier [Madame Bachelier is a fictional character from the story]...
From *Fictions* ed. A. Kerrigan Jupiter books 1965

Borges here is ironically overstating an important truth about the reading process. Our preconceptions about when a work was written and by whom are always important in our experience and judgement of it, whether we are conscious of this or not. So deliberate misdating at least helps us to be conscious of these powerful preconceptions. We react to *Samson* as a neoclassical work, by a 17th-century Englishman trying to recreate a Greek tradition of the 5th century BC. How would we look at it if we discovered that he had really copied it out from an old manuscript, and passed it off as his own? Or take the Latin poems. They are usually regarded as derivative performances. But Coleridge anticipates Borges when he writes: 'You will not persuade me that, if these poems had come to us as written in the age of Tiberius, we should not have considered them to be very beautiful' (*Table Talk* 23 Oct 1833).

And even if M had simply copied out poems by Ovid, that would have been a significant act for a puritan like M, a defiant resistance against the harsh values he had absorbed from his background. And to be less fanciful, for him to publish in 1645 and then in 1673 what he had written in his teens and early twenties was to reaffirm, or at least not repudiate, that earlier work. How extensively did he revise for the 1645 edition? The technique of deliberate anachronism is worth trying extensively for the new perspectives it gives on works, which in turn bring out new experiences of the same poem, new poems.

Literary slumming

Milton wrote most of his early poetry in Latin. Nowadays this reduces their likely audience drastically, and makes him sound very bookish – as he was. But Latin also provided an escape for the young M, an escape into adolescence. Latin was the language used by Ovid and Propertius: legitimate soft pornography for an emotionally deprived youth. Later in life M referred to the smooth elegiac poets, whereof the schools are not scarce, whom both for the pleasing sound of their numerous writing, which in imitation I found most easy and agreeable to nature's part in me, and for their matter, which what it is, there be few who know not, I was so allured to read that no recreation came to me better welcome. *Apology for Smectymnuus* 1642

M was 33 when he wrote this in 1642, just prior to his marriage. Note the embarrassed reticence of 'which what it is, there be few who know not'. But he didn't have to mention it at all.

17th-century assumptions made it difficult to assert the value of this kind of literature, or insist on the special imaginative needs of children. We owe the definitive recognition of childhood to the Romantics. Here is part of Wordsworth's defence of children's taste in literature:

> O give us once again the wishing-cap
> Of Fortunatus, and the invisible coat
> Of Jack the Giant-killer, Robin Hood,
> And Sabra in the forest with St George;
> The child, whose love is there, at least, doth reap
> One precious gain, that he forgets himself...
> Dumb yearnings, hidden appetites, are ours,
> And *they must* have their food.
>
> *Prelude* v 341, 506 (and see the whole book)

M's sources of such 'food' were probably more limited than Wordsworth's partly because the young M was so much more earnest, and needed this kind of literature to seem somehow respectable before he could acknowledge it. The schizoid relationship between rich fantasy and accredited learning used as a cover can be seen at its clearest in his *Vacation exercise* 59–80. The content is the driest of philosophy, the doctrine of Substance, but the form is a fairy story, complete with fairy godmother, prophecies at birth and an invisible coat.

In the early poetry, especially the Latin poetry, the forbidden reading which nourished M's 'dumb yearnings and hidden appetites' is closer to the surface: but look for it in M's mature verse, too. There is an art of subversive reading which M may have used, which transmutes the classics (respectable works that can be read in the class-room) into something like their analogous popular form. Epics like Homer and Virgil contained stories of incredible valour, which might nowadays be found in war films or stories. The romantic epic of Ariosto and Spenser covered the full range – fighting, magic, sex. Even the Bible has some good stories – like Samson. What has the myth of Samson got in common, for instance, with the modern Superman? Some common elements: superhuman strength; a single source of weakness – hair/Kryptonite; total patriotism, to Israel/the American way of life; passivity before a dominant female – Da lila/Lois Lane; plus the opposition, blindness/X-ray vision.

What distinguishes great literature from popular literature? Is it a greater intelligence, complexity, self-criticism in the use of the common materials? Here is a Beatles song:

> Who knows how long I've loved you
> You know I love you still
> Will I wait a lonely lifetime
> If you want me to – I will
>
> For if I ever saw you
> I didn't catch your name
> But it never really mattered
> I will always feel the same
> (From album *The Beatles* Northern Songs 1968)

The song seems a nothing, an incoherent tissue of clichés from other pop songs. This is precisely the point of the song. 'If I ever saw you' is the give-away. No actual girl has generated the song. 'You' is the creation of the genre, perhaps fleetingly embodied when an actual girl briefly coincides with the mood of such songs, and seems to represent the ideal. The clichés of loving forever ought to be totally undercut, since they are seen to be attached to nothing, but MacCartney's voice is without irony, wholly endorsing the validity of the emotion even when deprived of its apparent object. M's *Elegy VII* works similarly. There too the emotional charge is set off as much by the Ovidian tradition as by any external object, but is not less real for that. Did any particular girl exist? M too didn't catch her name, but did it ever really matter?

Further reading
On the official curriculum, and M's reading within it, the fullest treatment is H. F. Fletcher *The intellectual development of JM* 2 vols 1956–61. To experience the vigour and subversive content of some of M's reading, try Ovid (*Amores*, in a very good modern trans. by Guy Lee, 1968; *Metamorphoses* trans. H. Gregory, Mentor 1958); Propertius (trans. Musker 1972); Aristophanes (*Frogs* trans. Arrowsmith or Dickinson: watch out for bowdlerized translations of Aristophanes, who received rough treatment by Victorian translators). On youth culture as a counter-culture in

modern society, see L. Langman 'Dionysius – child of tomorrow: notes on post-industrial youth' in D. Gottlieb ed. *Youth in contemporary society* Berkeley 1971.

Milton and Latin

Latin was not only a cloak for M's reading, the equivalent of plain wrapping. The forms of the Latin language had a decisive effect on his own style, for good or ill. Here is Leavis' famous condemnation of M's latinate diction:

> The extreme and consistent remoteness of Milton's medium from any English that was ever spoken is an immediately relevant consideration. It became, of course, habitual to him; but habituation could not sensitize a medium so cut off from speech – speech that belongs to the emotional and sensory texture of actual living and is in resonance with the nervous system: it could only confirm an impoverishment of the sensibility... So complete, and so mechanically habitual, is Milton's departure from the English order, structure and accentuation that he often produces passages that have to be read through several times before one can see how they go, though the Miltonic mind has nothing to offer that could justify obscurity – no obscurity was intended; it is merely that Milton has forgotten the English language.   *Revaluation* 1936

This of the author of *Hail, native language*. The oppositions here, spoken versus written, private versus public, natural versus acquired, native versus foreign, were all important and problematic ones for M himself: the blind and learned poet, connected to the world only by human voices, intensely patriotic despiser of his countrymen. Leavis' criticism of M is subsumed in M's own struggle with himself.

The remoteness of Latin no doubt attracted M. A different language is a different world, and M was always fascinated by different worlds. To repudiate one's own language is to repudiate one's own society which is constituted by that language. M was conscious of doing this, particularly in his Italian poems (eg *Sonnet 3* and his *Canzone*). Sometimes he seems merely snobbish. William Marshall engraved a portrait of M for the 1645 *Poems* which M felt was unflattering. His revenge was to write an epigram in Greek, which Marshall could not understand, then get Marshall to engrave it under the offending picture. The epigram invited M's 'friends' to laugh at the unskilful artist: only readers of Greek, a scholarly elite of Europe, to be included as M's friends.

7

But to enter the world of another language, another culture, is also to gain a vantage point on one's own, to see new possibilities of thought and feeling. That is why this edition uses facing-page translations. No translation could be as alien to English as Latin is. Use the opportunity of seeing how untranslatable the Latin is, what things can be done easily in Latin which are difficult or impossible in English. To take an example that is very ordinary as poetry, the first line of *Elegy VII*:

Nondum	blanda	tuas	leges	Amathusia	noram
not yet	lovely	your	laws	Amathusian	I knew

First the possibilities opened up by Latin's freer word-order. In the experience of the line there is a typical suspension of meaning till the end. *noram* which fixes the relation of all the other elements to each other is delayed till the end: 'I knew' not known as long as possible. Similarly *blanda* waits for *Amathusia*, surrounding and colouring *tuas leges*, so that before we get to 'laws', which suggests something harsh and restrictive, we know that the lawgiver is gentle, attractive and female. (Until the last syllable, she could have been the subject of the verb.)

There are other ways in which this is alien to English. *Amathusia* for instance. This is really the goddess Venus, referred to only as a place, the feminity indicated only through the ending in *a* which shows that the word is feminine. But that is only the last of the transformations that have taken place. The underlying meaning would emerge in English as something like 'I had not yet fallen in love', but only 'I' and 'not yet' have a close equivalent in the Latin. There, a problem of feeling is presented as a problem of knowledge: knowledge about geography, theology and politics rather than psychology. The surface of the language is about an eroticized geography, combining a Miltonic sense of huge spaces with a comforting immediacy of relationships, since he is supposedly addressing this goddess of Amathusia with the intimate *tu*. At a deeper level, it uses a kind of psychological model very different from the puritan model M was brought up on. In this classical model, the emotions are external, hostile powers, divine in status but cruel and immoral and nothing to do with one's real self. Love becomes comprehensible as an experience of an extremely arbitrary kind of government, which makes love itself a specifically political act. Compare all this with what is suggested by modern English idioms: 'fall in love'.

Translation

The translations in this edition are meant to be fairly close and literal, to enable you to dispense with them as soon as possibel. That inevitably makes them unsatisfactory as translations from other points of view. They are not poetry in their own right. But all translations are bound to be unsatisfactory compromises in some way, either missing out on qualities of texture for the sake of accuracy, or losing accuracy in the interests of poetic effect. Some people believe that a translation should make good sense in its own right, the best translation being a kind of parallel creation, good poetry or prose in the new language. Against this is the view of Pannwitz:

> The basic error of the translator is that he preserves the state in which his language happens to be instead of allowing his language to be powerfully affected by the foreign tongue. Particularly when translating from a language very remote from his own he must go back to primal elements of language itself and penetrate to the point where work, image and tone converge. He must expand and deepen his language by means of the foreign language.
>
> From G. STEINER *After Babel* OUP 1975

This is close to how M must have proceeded – to Leavis' disgust. Both kinds of translation are immensely creative. In one a new work is created, in the other, a new language. Both are also equivalently destructive, of the original work or of one's native language. Translation involves criticism, and choices based on that criticism. Everyone should reach the stage of being dissatisfied with the present and every other translation of M's Latin poems, and use whatever aids are necessary to offer translations of their own, preferably several kinds. Eg aim for lucidity and elegance: or for a tortuousness equivalent to Milton's: or focus on and translate only the parts of the text where you think the real poetic impulse is to be located (which may be only one or two words per line).

Further reading

Use the largest dictionary you can find: Lewis and Short, or the *Oxford Latin Dictionary*, which is slowly working through the alphabet. They set out the possible meanings for a word clearly, and from there you can guess as well as any scholar which ones might have been stirring in M's mind, and could be relevant for a particular context.

Commentary on the shorter poems

Friendship: Elegy I

Elegies I and *VI* and one of the Italian sonnets were written to Charles Diodati. *Elegies V* and *VII* and the rest of the Italian sonnets probably were written for him, too, and *Epitaph for Damon* commemorated his death. There is no doubting the importance of the friendship to M. But what was the exact nature of that relationship? What was its place in the nexus of relationships which surrounded the young M?

The picture we have of him is as a lonely child: 'When he went to school, when he was very young he studied very hard and sat up very late, commonly till 12 or one o'clock at night, and his father ordered the maid to sit up for him', wrote Aubrey, an early biographer. M reported this routine himself, but characteristically didn't mention the maid. What was her function? Mother/servant? Did he notice her? The habit of not registering the existence of the majority of those around him, especially servants, persisted throughout his life. So in *Elegy I* he does not mention his father's household servants, and servants are even less to be seen in his world than in Jane Austen's.

It is in fact a striking absence of society that distinguishes *Elegy I*. Even his father only exists as an adjective attached to his house. Was his mother there? Probably. His brother? If so he isn't mentioned. The relationship with Diodati, several hundred miles away, is the only particularized relationship in the poem. Why Charles? Did M have no other friends? They had been schoolfriends at St Paul's, Charles two years younger, but apparently more precocious, since he went to Oxford at the age of 13 (M went to Cambridge at 16). Their ways seemed to have diverged, yet there is no other friend from school or university whose existence is registered like Diodati's.

But compensating for the absence of real relationships is the fictional society of the poem, the endless succession of characters who pour from the pages of his books or out of an actual or imagined theatre. These characters may have been from plays he saw in London, but they seem suspiciously literary, the products of his solitary imagination rather than the creations of actual people, the actors, in a social context, a theatre.

Even Diodati only intermittently seems real. The poem is addressed to him, but the poem is full of apostrophes (turnings aside, a rhetorical device), where he addresses, among others, Virgil, Adonis, heroines from ancient cultures, foreign women, and the city of London. This highly literary device creates a shadowy audience for M's performance, an audience completely controlled by the poet, appearing and disappearing at his will.

Two letters from Diodati to M survive, both in Greek. One is usually thought to have been the one that triggered off M's elegy. (What would make one friend write in Greek, the other reply in Latin verse?)

I have no fault to find with my present mode of life, except that I am deprived of any mind fit to converse with...But you, wondrous youth, why do you despise the gifts of nature? Why do you persist inexcusably in tying yourself to your books night and day? Live, laugh, enjoy your youth and the present hour, and stop studying the enthusiasms, licenses and recreations of the wise men of old, meanwhile wearying yourself. I, in all things your inferior, both think myself and am superior in this, that I know a moderation in my labours.

In *Elegy I*, M seems to be reacting to some such criticism. Literature in lines 25–46 is abundant life, and he insists that he does look at the gifts of nature (ie pretty girls) too – though literary allusions and a patriotic ecstasy quickly prevent anything happening. The extravagant praise of Ovid over Homer and Virgil seems part of the same enterprise, to prove to Diodati that premature senility hasn't totally prevailed yet. Clearly Diodati was valuable to the young M as a critic of M's severer puritanical tendencies, urging a more generous and social form of life. Diodati here shows himself quite acute about M, catching the paradox of M *studying* the 'licenses and recreations of the wise men of old': M's earnestness in examining pleasure, his need to displace it onto virtuous men of the past. M's elegy doesn't take this point at all – if anything his performance confirms it.

The only solid news the letter contains is the mention of his temporary expulsion from Cambridge. Outside this elegy the evidence for this episode is slight and inconclusive. Aubrey's *Life*, which drew on M's brother Christopher for some of its information, said of M's Cambridge career: 'His 1st tutor there was Mr. Chapell; from whom receiving some unkindness (whip't him), he was afterwards (though it seemed opposite the rules of the college), transferred to the tuition of one Mr. Tovell.' If he was in fact sent down for a period (the evidence is inconclusive) it must have been distressing to a youth as isolated as M from his fellow-students, and anxious to please authoritative males. So the brief way he refers to it may be a sign of his deep reluctance to talk about it even to his close friend. But distance is a recurring quality in the poem. There is real geographic distance, from Chester to London, with the river Dee leading into the Irish sea, away from M and civilization (so really it's Charles who is exiled, not M); and the distance of London from Cambridge. Literature also provides innumerable places as well as people: Tomis, Susa, Nineveh, Rome, Greece, Troy, leading to the glowing vision of London, outside human space and time, one of M's many versions of the New Jerusalem. The ending sounds weak after this, with its relapse into real space and time: his proposed return to Cambridge for the start of the next term. The robust, satiric tone of this conclusion doesn't sound like a youth who has been crushed by institutional disapproval.

Further reading. D. L. Clark *JM at St Paul's school* 1948. D. C. Dorian *The English Diodatis* 1950.

Nature and love: Elegy V

Elegy V was an extraordinary poem for M to write: extravagant, self-indulgent, totally pagan, with no attempt to moralize his classical sources. He gave his age as 19. Who he wrote it for is not known, but only Diodati sounds a plausible recipient.

It seems to be a celebration of nature, but nature as the Romantics knew it has no place. There are no plants or fruits or flowers growing, just a riot of human sexuality. The deities M draws on from classical mythology derived from a more primitive system of belief in which human sexuality was seen as continuous with the energies of growth and fertility in the rest of nature. Does M still see this link in his use of these mythological figures? Or is classical mythology used as a pretext for a frank description of human eroticism, transposed onto nature then immediately embodied by very human deities? But even so, the sexuality exists at two removes from M, and the poet takes no part in the orgies he describes.

Greek and Roman writers used the figures of the gods as reference points in their thinking about problematic relationships, between man and nature, and different aspects of the self: especially the deities who presided over the emotions. To see inner conflict in these terms is to represent it as both inner and outer, a struggle of one's self against lower impulses which are also divine. Self-control then comes to seem blasphemous, futile and self-destructive, but also the only surviving moral imperative.

The most powerful classical statement of this dilemma comes in the plays of Euripides. In *Hippólytus*, the youth Hippolytus, worshipper of chastity, is destroyed by Aphroditè (Venus), who makes his stepmother Phaedra fall in love with him. In *Bacchae* Pentheus, self-contained rationalist, refuses to worship Dionysios (Bacchus) and is driven mad, tries to witness the orgies led by his mother, and is torn apart by her and her attendants. In the 19th century Nietzsche restated a profound polarity in the human spirit in terms of the Greek gods Apollo, god of light and of poetry, and Dionysios, god of wine and patron of drama: the twin spirits, as Nietzsche thought, of artistic creativity.

Apollo is at once the god of all plastic powers and the soothsaying god. He who is etymologically the 'lucent' one, the god of light, reigns also over the fair illusion of our inner world of fantasy...Dionysiac stirrings arise either through the influence of those narcotic potions of which all primitive races speak in their hymns, or through the powerful approach of spring, which penetrates with joy the whole frame of nature. So stirred, the individual forgets himself completely...Not only does the bond between man and man come to be forged once more by the magic of the Dionysiac rite, but nature itself, long alienated or subjugated, rises again to celebrate the reconciliation with her prodigal son, man. The earth offers its gifts voluntarily, and the savage beasts of mountain and desert approach in peace. The chariot of Dionysos is bedecked with flowers and garlands: panthers and tigers stride beneath his yoke... Now the slave emerges as a freeman: all the rigid, hostile walls which either necessity or despotism has erected between men are shattered. Now that the gospel of universal harmony is sounded, each individual becomes not only reconciled to his fellow but actually at one with him – as though the veil of Maya had been torn apart and there remained only shreds floating before the vision of mystical Oneness. Man now expresses himsel through song and dance as the member of a higher community: he has forgotten how to walk, how to speak, and is on the brink of taking wing as he dances.

Birth of tragedy 1872, trans F. Goffling,
Doubleday 1956

Apollo as the god of poetry figures large in M's poetry. In *Elegy V* it is Apollo who comes in his vision, but a most un-Nietzschean Apollo, an inciter of sexual energies, bridegroom of the earth. What Nietzsche saw as opposites, M here unites in a single figure. But this was a reconciliation which could not survive beyond this particular poem. Elsewhere M saw a similar antagonism to Nietzsche's. It continues under other names in *Samson*: Samson = Apollo, a Nazarite forbidden the use of wine. Dagon is repeatedly associated with feasts and wine. Samson (blind but not drunk) ends the feast in an orgy of destruction, a 'slave who emerges as a free-man', literally destroying all rigid and hostile walls. *Samson* is a drama (Dionysiac) not intended for performance (Apollonian).

1–28 Poet as seer

The poem is placed in time in the opening lines, a cyclic time in which winter gives way to spring, but spring and summer will in turn be followed by autumn and winter. So the apparently unqualified panegyric to spring is framed by a reminder that spring is only one part of the year's cycle. The earth is the first of many middle-aged mothers who recover their sexual potency: see also Venus in line 103, and Aurora in lines 49–52, who is looking for a younger husband.

The poet's energies return with spring, human creativity with natural growth: the poet briefly identified with nature, in preparation for his projection onto nature. (He doesn't mention noticing any sexual feelings in himself.) The vision of Apollo has a hint of a sexual origin in the mention of Daphne's laurel, but the ecstasy that follows quickly leaves the human world. It's a paradoxical movement. His own soul leaves his body far behind, though the poem is a celebration of physical energies. No-one else's mind leaves his/her body. His initial ecstasy seems to leave the poet safe, uncontaminated by the orgies he then witnesses.

But the movement away from the body is also a movement into body: up becomes down. The pure liquid air becomes clouds, then shadows, then caves, and he continues inwards, peeping at forbidden mysteries, the shrines of the gods of heaven and hell, life and death are equally visible to this poet–magus.

14

He has passed beyond good and evil, and finally does not even know what he is saying in his ecstatic state.

The last reference to the poet's sense of his role, however, is much cooler. Philomela is to sing in the country while M sings in the city. So M sees himself as essentially an urban poet, in spite of the praise of nature. Philomela the nightingale is puzzling in other ways. As the victim of an act of rape, she is not the obvious celebrator of the orgies to follow. The mood here seems very different from the Dionysiac–Apollonian rapture we've just seen: one of a number of such shifts of key in the poem.

29–94 The wooing of the Earth

The celebration of spring begins astronomically, coolly describing celestial concomitants of spring. Some of the details seem thematically appropriate: eg the sun leaving the fields of Tithonus, the East (cf lines 49–54, where it is the female Dawn who leaves Tithonus). Other details are less obviously relevant, like the hints of a war in heaven. M's early poetry has a number of allusions to rebellion which seem irrelevant both to their context in the poem and to immediate historical circumstances, but foreshadow later events and major Miltonic themes. On this occasion the gods have no good reason to suppose that 'force' has withdrawn: see the riots that are to follow (though of course it is the gods now who are doing the forcing).

Equally hard to integrate into the poem is the Apollo of lines 43–4, who seems to be sexually inactive at night now, or so the shepherd believes. His virginal sister is pleased by his abstinence and new efficiency. Apollo then becomes the inciter of Dawn to adultery, and then himself switches from active to passive, from pandar to the object of mother Earth's attentions. The poet now is the pandar or marriage-broker, addressing Apollo on behalf of the Mother, his identity slowly merging with that of 'lascivious Earth'. She repeats to Apollo what the poet had said on her behalf, but her tone is very different, with its sense of coolness, evening, rest, and the ambiguous refrain: lay your brightness/eyes in my lap. Look? Make love? Die? Agave, mother of Pentheus, woke with the decapitated head of her son in her lap in Euripides' *Bacchae*.

94–140 Bacchanal

In the wild orgy that follows, there is no pain, no unpleasant consequences. Rape is impossible, because all virgins consent.

But with line 105 a more moral note briefly enters. Hymen the god of marriage makes everything legal and decent, every marriage a love-match, with no hints of jealousies or rejection. The problems of anarchy are dissolved away by a dream-like switch (in Nietzsche's terms, Dionysios becomes Apollo) and chaos becomes order. The antagonisms of real society are magically removed in the on-rushing illusion, with parents (mothers) making love to their sons, and Jove, the only member of a ruling class to be seen, giving a banquet for his servants.

The Dionysiac satyrs who follow, however, are almost comic in their promiscuity. Again, mothers are particular targets (Cybele and Ceres). Even this rape is not intended to bear fruit. The hectic sexuality seems to be its own justification. M offers all this as a return to the golden age, a reason for Jupiter to return to earth.

Dionysiac orgies are a common theme in baroque art of the 17th century. See especially Poussin (1593–1665), three *Bacchanals* in the National Gallery, London, and a bacchanal in front of the golden calf, in *Adoration of the golden calf*; and Rubens (1577–1640) *Drunken Silenus supported by satyrs*, also in the National Gallery. Rubens' satyrs are vigorous though decadent men, grappling with fleshy, matronly nymphs: orgies for the middle-aged. Falstaff is a satyr-figure too, and another anti-father. Hamlet compared his father to Claudius his step-father 'as Hyperion to a satyr' – again the satyr used to represent parental sexuality.

Eros and the poet

Elegy VII

This is a poem of considerable intelligence and honesty, arguably M's most subtle and successful venture at self-exploration in his early poetry. That makes its date important, so that we can place its insights and commitments in relation to the other early poems to which it is related. M gave his age when he composed it as *undevigesimo*, 19, a year earlier than *Elegy V*. But he numbered it *VII*, the last of his elegies, so either he regarded it as his most mature statement, or it was. (*Undevigesimo* is fairly close to *uno & vigesimo*, 21, so there is the possibility of a printer's error.)

It is certainly later than *Elegy I*, and its narrative touches on and goes beyond the early poem. There, he looked at some girls, but as a vague glowing mass, and he says he has remained untouched. In *VII* one girl emerges from the mass, and then departs: a brief contact with love which the poet finds more disturbing than *Elegy I* had suggested. The similarity of image and theme is so great that Diodati sounds the likely recipient of this poem, too. It's a Diodati-theme. The relation with *Elegy V* is less clear-cut. Apollo is important in this poem too, but which seems the earlier? The Apollo of *V* has Daphne's laurel round his brow, and is totally eroticized. The Apollo of this poem has to learn about Cupid's power. So either the lesson so painfully learnt in *VII* is taken for granted in *V*, or *VII* goes back to confront problems that were resolved by mere energy in *V*.

The starting point for the poem seems to have been a story from Ovid's *Metamorphoses*, Apollo's confrontation with Cupid. Ovid's story is repeated by Cupid in lines 31–4 of this poem, and its plot structures the whole of the elegy. Ovid's Apollo challenged Cupid, as M does at the beginning of the poem. Cupid's revenge was to make Apollo fall in love with Daphne, and make her reject him: M's case also. So M is dramatizing Ovid's story, seeing himself as Apollo. But the details have biographical point, and suggest greater self-criticism than any of the other earlier poems. For instance, in *Elegy I* he ducked Diodati's criticism of his bookishness. Now (lines 15–16) he admits it, and gives the realistic and symbolic detail of his eyes blinking, dazzled at the light of day. Like Blake's *Mad song*.

> I turn my back from the east,
> Whence my comforts have increased.

The image of Cupid appearing beside his bed is conventional in detail, but this Cupid is also an attractive boy, compared to two famous pretty boys from antiquity, Ganymede, beloved of Jove, and Hylas, boyfriend of Hercules. Is M suggesting early homosexual attraction (not of course any homosexual acts)? M's nickname at university was 'the Lady of Christ's'. At university, boys slept two or three to a room, often with a tutor.

Cupid's list of conquests suggests the destructive nature of love. Apollo is the first, exemplary victim: arrogant, self-confident, a public hero who slew the monster Pytho – like Samson. This is far from the Apollo of *Elegy V*. Then follows the 'cretan hunter', perhaps Hippolytus, another self-sufficient worshipper of chastity; then Cephalus, who appeared in *Elegy V*, too. The difference between the two poems can be seen in their treatment of Cephalus. *Elegy VII* stresses the crime he committed, accidentally murdering his loving wife, the personally disastrous consequences of love, whereas *Elegy V* filters out the complications of Dawn's husband and Cephalus' wife. Cupid's speech ends where it began, with Apollo, who will be no use to M.

The erotic experience that follows is innocent enough, a matter of eyes touching, and a growing interest in one girl. Cupid aims his arrows discreetly in this poem: eyebrows, cheeks, lips, nothing below the neck. Apollo's interest was less restricted in Ovid:

> He marvels at her fingers and hands
> and arms and upper arms bare to the shoulder,
> and what is hid he thinks still lovelier.
>
> *Met* 1 500

M is so preoccupied with the intensity of his response that the girl disappears while he is still overwhelmed by the new sensations. This is the same self-concern as in *Elegy I*, but it is placed in a world where people who are ignored are liable just to walk off.

The sense of loss comes over powerfully, with suggestions of blindness and damnation, foreshadowing both Satan and Samson. The end of the poem then formally acknowledges the power of love, but as an amoral, irresistible force, not even a relationship. So if he is to fall in love again, he prays, let it be mutual, not the frustrated and destructive love of an Apollo-figure.

Elegy VI

This poem is probably the last of the elegies to be written, probably from the same time as *Nativity ode*, when M was just 21. It is addressed to Diodati again, and contrasts the two as two kinds of poets. Diodati explicitly unites Dionysios (Bacchus) and Apollo:

> Song is a lover of Bacchus, and Bacchus of song;
> Apollo thought it no shame to wear clusters of green,
> and valued the ivy of Bacchus above his laurel.   14

This is the Nietzschean opposition resolved by Diodati, but M sets a different aim for himself, as poet–priest:

> The poet of wars and the heaven of ageing Jove,
> and pious heroes, and leaders half divine,

who sings of the solemn decrees of the gods above,
    and the realms below, where fierce Cerberus howls,
must live a frugal life, as Pythagoras did,
    wise teacher, feeding on simple herbs.
Beside him a bowl of beech with crystal water,
    drinking sober draughts from a pure stream.
And his youth must abstain from evil, chaste his ways,
    morally strict, his hand without a stain.
Like you, O priest, resplendent in gleaming robes
    and holy water, placating the hostile gods:
Like this the wise Tiresias lived, they say,
    when he lost his eyes, and Linus of Thebes,
and Calchas iń exile, his house destroyed, and aged
    Orpheus, who tamed wild beasts in lonely caves.
And Homer, who ate small meals, and drank from streams...55-71

The poem reads like a correction of *Elegy V*. The poet who was the magus of *V*, with a vision of heaven and hell, continues that task, but with greater solemnity and dedication to militaristic values, no longer the cleebrator of the rites of spring. *Semideosque duces* (leaders half-divine) sounds like a deliberate echo of the striking *semideusque caper* of *V*: the grotesque union of god and goat replaced by the noble union of leader and god. The reconciliation of Apollo and Cupid of *VII* is also repudiated, and normal social life rejected in favour of a regimen like Samson's (cf 541-52, 533-40). M, aged 21, identifies with old men, two of them blind (Teiresius and Homer): an ominous commitment; see *PL* III 35.

Epilogue to the elegies
These lines carry the rejection of Eros even further. They seem to have been written long after the rest, perhaps just before publication in 1645. He now sees through his earlier industrious self-indulgence. The imagery is savage, close to the grotesque in its exaggeration. Do the painful images suggest his sense that this decision is wrong in human terms? It's a sombre, wounded poem. It would be interesting to know whether it was written before or after Diodati's death, before or after his marriage and separation from Mary Powell.

Translation from Horace
This is M at his finest as a translator. The poem exists somewhere in a linguistic hinterland between Latin and English: tortuous almost to the point of incomprehensibility, sometimes using the Latin root meaning rather than the English sense of

words, more compressed than English allows, moving through the stanza form as though it didn't exist. Paradoxically, M manages to be more unlike Horace because of his closeness to Horace's language. A translator like Michie, in the Penguin classics, tries to recreate Horace's elegance and urbanity, but inevitably takes twice as many words as Horace to do it as he attempts all the time to write intelligible English. M's poem is an assault on the English language, only half as many words more than Horace, communicating a sense of struggle, effort, suppression and dislocation of energies: quite unlike Horace's in tone. Horace is wry in the warning he gives the youth, self-mocking, giving thanks to the 'powerful' god of the sea for saving him. M changed 'powerful' (*potenti*) to 'stern', to give the repudiation of oceanic sexuality a more sombre and definitive tone.

It is not known when M wrote this. The Horace poem follows a Horatian celebration of spring which is similar in commitments to *Elegy V*.

May morning
This is a slight poem, a fragment of *Elegy V* reworked in English. In the 1645 edition it follows *Winchester*. If this fixes the date, it was written May 1631, a year after *Elegy VI*, two years after *V*, showing M returning again to that poem to exorcize the energies it had released. Its tone is lucid, organization unobtrusively adroit, images discreet – and, of course, it's in English, a particularly clear and natural English for M, with run-on lines and normal word-order. The apparent simplicity goes with careful art. The rhyme-words, for instance, are all different parts of speech, noun rhyming with pronoun, verb with noun etc. Nouns and adjectives go in pairs: Venus and May, *flowery* May and *green* lap, yellow cowslip and pale primrose (primroses are yellow too). The pairs typically are couples, not opposites, their relationship harmonious, a linking of arms carried along by the fluent movement of the verse: an image of beneficently ordered energies, not the orgy of *Elegy V*. Politically it is Anglo-catholic, if M was conscious of such things then. Strict puritans disapproved of May Day festivals.

Italian sonnets

The Italian poems were probably written to Diodati, who was of Italian extraction. They occupy an equivocal place in M's poetic development. They are written in a foreign language, one that is not very foreign (compared for example to Greek), but M is conscious of its difficulty. They also continue M's attempt to reconcile sex and morality, drawing on a different poetic tradition that had been concerned with precisely this conflict for centuries (from Dante through Petrarch to della Casa in the 16c, all poets greatly admired by M). In an idealized biography of M these sonnets come after his early pagan enthusiasm for Ovid. Where they come in his actual biography is less sure. The more favoured date is 1630, when M was 21. This would have been at the same time as *Elegy VI*, with its apparent rejection of amatory themes and its commitment to epic: hence these sonnets reject specifically that rejection. Some scholars date the sequence in 1638, the period of his Italian journey. Where does he seem to be? England, or Italy? Stylistically the poems are an advance. Influenced by the style of della Casa in particular, M wrote with a new freedom and complexity, with syntax sometimes difficult and obscure.

Sonnet 2 Donna leggiadra

At first sight this seems a conventional exercise in flattery. The lady's name is Emilia, conveyed by an elaborate code in the first two lines, and it's really her virtue that M loves. This was the standard resolution of the conflict between love and virtue for the sonnet form. But M's treatment of the theme is more interesting and intelligent than this summary suggests. The *nobil varco* is the Rubicon, the famous river Caesar paused at in his advance on Rome. 'To cross the Rubicon' came to refer to a decision from which there was no turning back. The poem moves through a complex network of either-ors. The first is either to love or not to love, with the non-lover disvalued. But this choice turns out to be an illusion, because everyone must love her. The alternative then becomes whether the lover is worthy or not. If he is not worthy, then there are two more possibilities. Either he will be saved from loving by grace from above, or he will be permanently in love. The complex situation can be represented schematically like this:

```
⎧ worthy lover (pure, regenerate) → unrestrained, virtuous love
⎨                                  ⎧ no grace (inveterate, sinful desire)
⎩ unworthy lover (fallen man)    → ⎨
                                   ⎩ grace (celibacy)
```

From the point of view of the lover, the situation is very ambiguous. If M loves Emilia, for instance, he might be entirely virtuous, or worthless and lost in desire: very good or very bad. If he doesn't love her, he has been totally lacking in virtue, but has been saved by grace. This makes the austere celibate of *Elegy VI* less admirable than a virtuous and passionate lover, but more admirable than a merely passionate lover. If Emilia isn't completely virtuous, the position is even more hopelessly tangled. These doubts, which correspond to the important theological dispute as to whether fallen man has been restored to his original purity by Christ's saving grace, make this poem a strenuous and troubled work, more than the conventional tribute it seems at first.

Sonnet 3 *Qual in colle aspro*
This poem also shifts disconcertingly into theology in the last two lines, showing considerable tension between love and morality. The first four lines present a stylized picture, which M interprets in two senses. The first interpretation is unusual and strained, to do with his learning of Italian. The images are gross, the vanity naive, not much of a compliment to the girl. But he finishes with a second reading of the allegorical picture, in which heavenly love replaces secular love. This second reading is prepared for by the original image, where the mountain was suitably barren, and the 'strange plant' from a distant country suggests a common metaphor for the hostility of earth as a site for the growth of heavenly virtues (cf *Lyc* 78: 'Fame is no plant that grows on mortal soil'). His praise of his linguistic competence still feels like an intrusion, jarring with the delicacy of the opening scene. Without it, the poem would have a commonplace air of melancholy piety. With it, it has the interest and structural defects of a poem about to split in two.

Canzone
This poem starts ostensibly with a jostling world of courtly lovers, but they remain shadowy presences, in what sounds like a private dialogue continuing themes raised in *Elegy VI*. First they challenge him with using a language 'remote and strange'. Following the previous poem it's natural to assume that this is Italian, but this might not be so. Unknown – to whom? Italian was strange to some Englishmen, but so was Latin. English was strange to other European nations. M spent many years brooding

over the choice of the right language for his poetic ventures, particularly between Latin, the universal but dead language, and his own vernacular. Then the lovers 'joke', challenging him with the high conception of the poet's role which he offered to Diodati in *Elegy VI*. He has no answer to this, as it's his own ideal, but the last three lines refuse to renounce his commitment, to the lady and hence to love itself: a commitment stronger for being unjustifiable in any terms he can bring forward.

Sonnet 4 Diodati
This is *Elegy VII* translated into Italian and addressed to his friend. The details sound authentic, not only the dark hair but also the mastery of several languages: a suitably erudite girlfriend for M. (He didn't think it necessary to teach his own daughters foreign languages though.) The final image, of wax melting in the ears, sounds gross. Was it meant as a joke?

Sonnet 5 Per certo
If the final image of *Sonnet 4* might be a joke, all but the last image of this one surely must be, a heavy parody of the stock phrases of the convention. The principle is the analogy between the body and the earth. M usually attributed human energies to parts of an animated landscape. Here he works in reverse, treating bodily functions as geophysical events. The effect is to magnify grotesquely the symptoms of the lover, absurdly because he is the passive lover of the sonnet convention. It makes the emotions feel completely different from what most immediately expresses them, as though the lover's body not only doesn't belong to him, but is literally a different world: an experience of extreme alienation, disturbing as well as ridiculous. But M again changes direction abruptly, to finish with the beautiful lyricism of the dawn coming 'brimming with roses'. The image is as conventional as the others, but has its effect because of the inversion of scale: instead of sighs like geysers, the sky at dawn is small and human-sized as a girl with flowers.

Poetry and language

At a vacation exercise: 'Hail, native language'
This was probably M's first poetic manifesto, an important event for him. The occasion was a college assembly at the end of term

before the long vacation, probably in 1628, when M was 19. The conception of the role of the poet is similar to that in *Elegy V.* What is especially interesting in the *Vacation exercise* is the praise of English delivered in English. The normal language for such occasions was Latin. M had already delivered a Latin speech, preserved as his *Sixth Prolusion* (trans. P. Tillyard in *Milton, private correspondence and academic exercise* Cambridge, 1932). The Latin is vigorous, bawdy and abusive. Alongside it, M's English seems stiff, formal and inexpressive: but even to use English would have had its shock value.

The opening metaphor for language seems to be that of a childish alter ego, a kind of servant figure who operates the inefficient mechanism of his tongue, and was the real source of what the young M said. From line 15 he switches to another metaphor, the very common 17c metaphor of language as clothing to thought. He criticizes mediocre fellow-poets of his day in these terms, for their 'new-fangled toys' and 'trimming slight', which he contrasts with 'richest robes, and gayest attire', which are still kinds of covering, distinguished only by being more ornate and expensive.

The metaphor had roots in contemporary social practice. 17c society was organized in terms of status, which was signalled in a number of ways, one of which was clothes, a convenient pretext for a conspicuous display of wealth. M's sole concern in this poem seems to be not with sincerity (which would be nakedness) but richness, opulence; to give his thoughts maximum possible status. This theory would lead to language designed merely to amaze and impress with its visible guarantees of intellectual wealth – criticism that has been levelled at M.

When M goes on to announce his 'graver subject' the metaphor gets even closer to the financial. Language must search his/her 'coffers'. He uses the metaphor of money in his Latin poem to his father, protesting

> This page displays the extent of all my goods,
> I've counted all my wealth on this single sheet,
> for I own nothing but Clio's golden gifts.
>
> *Ad patrem* 12

For intellectuals in Elizabethan and Stuart times, something like this would have been precisely true, if they had no other financial backing. Their skill in language was their only saleable commodity, a commodity which needed to seem as impressive as possible. M's father had amassed sufficient wealth to guarantee M a secure living, but he had also invested capital in M's education, and had a right to expect a return, in the form of tangibly 'rich' language.

Ode to Rouse

This is M's last significant poem in Latin. In form it is ambitious, a Pindaric ode (a difficult Greek form transposed into Latin by an Englishman) but handled with masterly control and elegance by M, deft in its touch yet elevated in theme, qualities M usually found it difficult to combine. It was written in January 1647 for John Rouse, the librarian of the Bodleian Library at Oxford, to accompany a volume of his Latin poetry to replace a copy that had been mislaid. Oxford had been a royalist stronghold in the civil war, and had only fallen six months previously. Fairfax, the parliamentarian general, had shown his respect for learning by putting a garrison round the Bodleian to protect it from pillages.

The first strophe is assured self-criticism, not the assault on his earlier self of the epilogue to the elegies. First he notes the over-earnestness that never quite came together in achieved poetry. The innocent isolationism of the early poetry becomes the conscious policy of the embittered middle-aged poet. The ideal of poetry transcending its particular time and place shows its roots in an intense class-hatred.

He also notes the innocent asocial self-concern of his early poetry, but the image that closes the strophe, of his 'feet hardly touching the ground' has a certain ambivalence. It's a less extravagant version of the poet as visionary/astronaut, but it could also suggest a lack of contact with life.

In the second strophe he locates poetry firmly in its political context. The need now is for a saviour, a holy one (*sanctus*), who will be an aspect of Apollo (he uses Apollo's arrows) but is specifically a political force. Did he have Fairfax, the saviour of the Bodleian, particularly in mind? Two years later he wrote his sonnet to Fairfax seeing him as a Hercules-figure who might save England from the Hydra-monster. If M was working on *Samson* at this time, as some think, the saviour figure of *Rouse* would be a study towards the figure of Samson: a combination of Apollo and Hercules, surrounded by a depraved populace. The victim that is to be saved from the harpies, interestingly, is blind: Phineus, prophet–king who was struck blind by Zeus and afflicted by the harpies. By this time M knew he was going blind.

But even though this poem itself is an intervention in the current political situation, M ends it with a plea for the neutrality of poetry. He wants his book stored in the library, *Quo neque lingua procax vulgi penetrabit* (Where the insolent tongue of the rabble will not penetrate). Poetry cannot redeem the degenerate masses, so it must repudiate them.

Death

Death is universal to every culture, but the conditions of dying and the modes of coping with this supremely problematic and inexorable fact of life have varied from age to age and society to society. Strategies for coping with death are generally akin to important life-strategies, for the problems that death poses are recurring problems of life, but in a more acute form. Death of oneself is the ultimate in weakness, experienced in life in sickness, aging, humiliation, sleep, blindness, drunkenness, sex etc. All these areas of experience have always provided metaphors for death which are stronger, more intrinsic than metaphors, since no-one has ever experienced death, and will always be using it as a negative way of exploring or expressing an obsession with power. The death of others, bereavement, is quite different: the definitive loss of a beloved object, permanent discontinuity of a relationship.

The developed countries of the 20th century have achieved a quantitative change in the facts of death which seems almost a qualitative one as well. Owing to modern medicine very few infants die at birth. In the 17c nearly 30 per cent of children died in infancy. The shocking conjunction of birth and death was commonplace then. Now it seems more possible to keep death and life separate. The manner in which people die has changed, too. In the 17th century, death was a communal event. People died at home, surrounded by others, including children, going on with living. Nowadays it is more common to die in hospital, alone except for doctors, nurses and life-support systems, with perhaps some other terminal patients as well.

Gorer, in *Death, grief and mourning in contemporary Britain* 1965 found that attitudes to death vary with class. The professional middle-classes are particularly inept at facing death. Only one in eight is present at the death-bed of their loved ones. The custom has grown of not telling a dying man that his disease is incurable. In the 17th century people were reminded of death when they were not dying: now death is not mentioned when they are.

But Gorer also points to an important paradox. Death has not in fact been eliminated from our society, only delayed. So although the expression of grief is being eliminated, the reality remains as strong. It has merely been censored, and as a result is much more devastating in its effects. Other societies have had rituals for dealing with the distressing emotions aroused by death, as for other life-crises. The anthropologist Van Gennep has called these rituals 'rites of passage' in his book of that title (trans. 1960). These are rites that help to mediate an individual's passage from one condition or status to another, and assist society to adjust to the change. These rites commonly have three stages, which Van Gennep calls separation, transition, and incorporation. Rituals for the dead in these terms will typically separate the living from the dead, mourners from others, and prescribe a period of transition, after which the living return to the living, and the dead are fully incorporated into the world of the dead.

Beliefs about death also have an important function in controlling responses to death. The traditional Christian view is very consoling. It is dualistic in form: the body dies, but the soul lives on. Christianity also posits a resurrection of the body, a time when the body will revive and rejoin the soul, by a process that is not entirely clear. A belief along these lines is so comforting that many people today effectively hold it who are otherwise not practising Christians. There seems to be some consolation in repeating phrases which say something like this, when someone close to one dies. Paradoxically, M who was strongly committed to his form of Christianity did not accept the orthodox Christian account of death. He held what is called the mortalist heresy, that is, he believed that both soul and body die at the moment of death, but that both will be restored to life at the last day. This belief makes death a total break, but life for the resurrected saints a full-bodied existence.

There is a great danger of romanticizing the past in reaction against present nastiness. Ritual in the 17th century was distributed on a class basis. The poor died in their families, and the church provided a ceremony, but for the rich the ritual was fantastically elaborate and expensive. Jessica Mitford condemned the extravagance of the American death industry in her *American way of death* (1962). She estimated the cost of death to Americans in 1960 was two billion dollars, slightly more than was spent on the whole of higher education. She quotes from the *National Funeral Service Journal* of 1961: 'A funeral is not an occasion for a display of cheapness. It is, in fact, an opportunity for the display of a status symbol which, by bolstering family pride, does much to assuage grief.' Even in the 17th century the rich spent more on death than on higher education (there were only two universities, plus a number of law schools). The cost of a funeral for a member of the peerage was stupendous, and the motives of the family were probably exactly what American funeral directors had in mind:

Ostentatious in his lifetime in clothes and hospitality, transport and housing, in some respects an Elizabethan peer had his finest hour only after he was dead. So grandiose in scale and portentous in style were the funeral arrangements of the nobility that the most contemptible of human beings on earth could hardly fail to be ushered out of it to universal admiration. Of many could it be said that nothing became them so much as their going: it was the last tribute of a deferential society to the dignity of a title.

L. STONE *Crisis of the aristocracy* OUP 1967

But Stone notes a swing against Elizabethan excesses during the early 17th century, influenced by puritan objections. Even tomb monuments became less lavish, though they still involved considerable expense. Look at some 17th-century tomb monuments. The materials guarantee the wealth of the donors: marble and brass were beyond the reach of more than two-thirds of the populace. Part of the rituals included poetic tributes, which came in reams from large numbers of unemployed or aspiring intellectuals. M at the age of 17 wrote 4 poems in Latin for the deaths of minor notables, 2 bishops, a vice-chancellor and a beadle, none of whom he could have known well or cared about.

There are many materials to draw on, to build up a picture of modern reactions to death and loss. One place to start might be death notices in your local paper. Look for continuities, and discontinuities, with M and the 17th century. Eg what is the function of such notices in relation to the community? What would it imply for a family to put a death notice in a national newspaper? The bulk of death notices taken up with the names of the living, and their relationship to the dead person. The bereaved are the ones whose feelings really matter, who are given public status by their connection with death. Space is also taken up with places and dates. This is a modern obsession.

M never mentions places or dates. In modern capitalist society time is typically spatialized and made manageable through numbers: hence the apparatus of time-tables, time clocks, the near-universality of the wrist-watch. Precision about time and place, key to a modern individual's sense of control, is naturally extended to death, as though somehow it will help to know precisely where and when the death took place, where the funeral will leave from and go to. In the 17c this comforting sense of certainty was more likely to come from the church's precision about where the soul was going to go, and for how long.

The dates and times of modern funeral notices may also show a typical separation of the decisive moments: death in one place, life in another, and the ritual in a third. Over it all presides the undertaker from a fourth place that didn't exist in the 17c: the paid expert in death.

Death of a fair infant

This was M's first surviving poem in English. Edward Phillips, his nephew, said that it was written for Phillips' dead sister, who must therefore be Anne, born 1626, died 1628, when M was 19 (not 17, as M wrote on the published version, though one modern editor believes M). But why did he write it? He wrote nothing on the death of his sister Anne's husband, three years later. Anne was already the mother of one son, pregnant with another. M had lost two sisters in infancy himself. What was special about baby Anne?

The structure of the poem follows Van Gennep's stages: separation (stanzas I–IV, where she is lamented as dead); transition (stanzas V–IX, where M wonders whether she is truly dead, mortal or immortal); and incorporation (stanza X where baby Anne is in heaven as mediator, stanza XI where the mother is told 'cease to lament'). But other feelings cluster around the images, less clearly relevant to the task of comforting sister Anne. The opening line is erotic, as though for a girl much older. A two-year-old infant is hardly 'blown' yet. The image is of a rose, a common image for the passing beauty of woman. The normal context for this image was a *carpe diem* poem, advice to a reluctant maiden to make love while she is still young and beautiful:

> Look, Delia, how we esteem the half-blown rose
> The image of thy blush, and summer's honour!
> Whilst yet her tender bud doth undisclose
> That full of beauty Time bestows upon her.
>
> DANIEL *Sonnets to Delia* 1592

Baby Anne then becomes the victim of rape by Winter, an old rich man. Phrases like 'kiss/But killed', 'fatal bliss' come straight from an erotic tradition where the kiss was usually a real kiss, the 'death' a sexual languor, the 'dying' of coitus (though impotent old Winter doesn't 'die' on this occasion, only baby Anne does). Stanza IV introduces Apollo, M's favourite classical deity, lamenting a homosexual relationship. What would sister Anne have made of all this?

The second section begins with concrete images of death and decay of the body, which the poet says he doesn't believe in. Then he switches his attention upwards. Thus far the word 'soul' has only been mentioned once, in line 21 where M opted out of the sexual metaphor he had used till then. Now he refers to her consistently as 'soul 'or 'spirit', and it is as though her body has ceased to exist for him, as an object of revulsion or lust. Her real history has also disappeared and M's speculations attempt to set the phenomenon of her life in a cosmic drama whose outlines were to become familiar in M's later work: wars in heaven, the irruption of a spirit of sanctity onto a degenerate earth, an image of a final judgment. Baby Anne gets totally lost in all this, as does her mother.

The conclusion is more conventional. In stanza x baby Anne seems to become the Virgin Mary mediating for man in heaven – surprising and heretical in a puritan poet, though the puritanism hasn't been very prominent so far. Stanza XI begins with 'Then', but in spite of this doesn't follow on from what precedes. It is a self-contained pious exhortation, using the traditional language of moral coercion and bribery. Baby Anne is a 'present' to God (which wouldn't be the case if she was a star, Astraea, an angel or the Virgin Mary, as earlier) which He only 'lent' – a formula which is still common today in death-notices, funeral cards etc. Then comes the bribe. If you are patient now, God will give you an immortality of fame through your next child. Anne was then two months from giving birth to a girl, Elizabeth, who did not become famous.

Hobson poems
These poems, written in 1631 when M was 22, commemorate a death but there is no grief to be exorcized. Hobson was 86 when he died, a well-known figure to Cambridge undergraduates, who relied on his carrier service between Cambridge and London, but who were separated from him by impermeable barriers of

class and status. He was a worker, though a wealthy man when he died; they were intellectuals, future administrators, teachers, clerics, lawyers or poets. So the affection is real but patronizing, the jokes unfeeling, as though about someone who was only an honorary human, like a pet. And the jokes are addressed over the head of people like Hobson to other students (these poems were two in a collection of poems for the carrier). The needs of Hobson's relatives for comfort and consolation don't seem prominent in M's organization of his poems.

But the first of the two poems does have its strength, demonstrating a real respect for Hobson's vitality. The language is direct and colloquial, Anglo-Saxon, the style pitched at the right social level for its subject. The images personify death, avoiding all metaphysics, making the process seem as familiar and unproblematic as work. The images come from activities of a full and social kind: riding, wrestling, serving. Hobson at the end doesn't split into body and soul, as the fair infant had done. The final image of death as sleep is traditional but effective, because the whole poem has remained rooted in tradition, from the opening 'Here lies old Hobson'.

The second poem to Hobson is another matter. 'Another on the same' sums it up. The opening 'Here lieth one' forgets to name and commemorate Hobson, which was the point of the formula. The jokes, on the paradox that Hobson needed to remain active (no profound insight there) come from academic texts, and rely on polysyllabic language that would have excluded Hobson and all his class.

Epitaph on the Marchioness of Winchester

Jane Paulet, who died 15 April 1631, at first sight may seem a surprising subject for an epitaph by M, in the light of his later development. She was an aristocrat and catholic, her husband a royalist hero in the civil war, and it doesn't seem that M knew her personally. But performances like this were typical for someone in M's insecure position, as he was at Cambridge. A number of other poets, including the poet laureate, Ben Jonson, also wrote poems to commemorate her death. M was in good company.

Such poems often pretended intense personal grief (Jonson did in his poem for the marchioness): the poet thrusting himself into the ranks of the mourners, hoping to be noticed. M's tone is relatively restrained. But the function of such poetry, like the art of the American mortician, is to beautify the dead, and give the chief mourners a 'good memory picture' as it is called in the trade: to falsify the dead and flatter the living. The opening focuses on the tomb, 'rich marble' (contrast 'Here lies' for Hobson). The next two lines fix her status, as wife of a marquess, daughter of an earl. The 'virtues' alluded to after this are not specified any further, as though they can be taken for granted if an earl's daughter possesses them.

As with the 20th-century euphemisms 'the loved one...dearly beloved', the complex emotions of a real relationship are reduced to a two-dimensional image of virtue and love. Jane Paulet seems in fact to have been regarded as virtuous by her contemporaries (she was rumoured to be toying with the idea of turning protestant, which would have been real virtue as far as M was concerned). M gives a compressed biography for her, accurate in outline but misleading in some details. 'Her high birth, and her graces sweet Quickly found a lover meet': very quickly, since she married at the age of 14, like Shakespeare's Juliet, but to the arranged partner, not to any Romeo. 'Lover meet'? Jane probably did not have much say about that. He may in practice have been no worse than the average aristocratic husband in an arranged match. At a later age, M was to be critical of the institution of arranged marriages: here he tacitly accepts the whole system.

M continues the biography with an account of the birth of the first son, then her death and the death of her still-born child. In fact she died of an abscess in her cheek, which burst on being lanced. M suppressed this image, giving the marquis a better 'memory picture'. Or is the image of the 'pearls of dew' on the blossom a transformation of the streams of pus down her cheeks?

The poem is organized in 3 parts, following Van Gennep's divisions. The first part (to line 22) describes her life, full of portents of death. The second, lines 23–46, describes the death and funeral accompaniments. The third begins the ritual of allaying grief, addressing the lady for the first time. The climax of the poem celebrates her incorporation into the community of heaven. Heaven proves to be a class and status society, the celestial reflection of 17c England, except that the marchioness will be promoted to queen. M calls on the Bible here, with an allusion to Rachel, Jacob's wife. Did he remember that Rachel had had to share Jacob with another wife plus two concubines? Rachel would not have been a good example to choose for an

advocate of women's liberation or women's rights – but M is
certainly not that, here at least.

On Shakespeare

This poem is not strictly about death, but it was called an
epitaph when it was published in 1632, with the second folio
edition of Shakespeare's works. Shakespeare's death seems un-
problematically in the past, and his achievement is on the way
to immortality. The basic idea is the commonplace that a great
writer's best memorial is his work. M in pursuing the idea
suggests a critique of all tomb worship, the materialistic props
of the age, Egyptian death-worship as the ancestor of Jacobean
funerals and their modern equivalents.

But the poem finishes with a strange kind of celebration of
Shakespeare's power. As a death-ritual it works in reverse.
Shakespeare starts off dead and remote, but the poem incorporates
him with the living, who themselves become dead. This disturbing
mingling of living and dead is precisely what rituals are meant
to prevent. M is registering the power of the great writer to
paralyse his successors. Even two centuries later, Arnold was to
feel the same about Shakespeare:

> Foremost among models for the English writer stands Shakespeare: a
> name the greatest perhaps of all poetical names... I will venture, how-
> ever, to express a doubt, whether the influence of his works, excellent
> and fruitful for the readers of poetry, for the great majority, has been
> of unmixed advantage to the writers of it.
>
> Preface to *Poems* 1853

Ironically, T. S. Eliot said exactly the same of M's own influence
for 20th-century poets.

Epitaph for Damon

On 3 April 1637, M's mother died. On 10 August 1637, Edward
King, a fellow undergraduate at Cambridge, was drowned, and
M wrote *Lycidas*, a pastoral monody, to commemorate his death,
though M may not have known him well or liked him greatly.
But next year, in August 1638, Diodati died while M was travelling
in Italy. Two years later M finally produced *Epitaph for Damon*,
another pastoral lament. Comparisons between the two poems
are inevitable, and must have suggested themselves to M. The
comparison raises important questions about how poetry works.

33

Is sincerity essential to a great poem? Diodati's death certainly mattered to M: Edward King's probably didn't. But can a poet be too sincere? 'The poem almost seems to ramble, as if the author, half-benumbed by his emotions, did not quite know how either to stop or go on' (Bush *Variorum commentary* 1970). Then there is the problem of artifice. Bush again: 'In the *Epitaphium*, a few passages of poignant directness make most of the pastoralism stand out as artificial literary embroidery.' *Lycidas* has pastoral trappings too, but at least it is in English. But M elsewhere turned to Latin to express personal feelings more intense and private than he could contemplate in English. The point about pastoral artifice needs to be tested out on the poem.

Unlike all M's other poems on death, this is the poem of a mourner coming to terms with his own loss. The delay in writing it seems puzzling. After he heard the news of Diodati's death he continued his Italian trip. In Rome the next year he apparently saw the opera singer Leonora Bari, and wrote two short poems in Latin in praise of her. So he was not incapacitated by grief. But the delay in writing is part of the poem itself, where it is given as something that somehow contributes to the grief or to the need to express it. If M's grief did last two years, it must have been almost pathological, what 17th- and 20th-century psychiatry would call melancholia. Here is Freud on the relation between mourning and melancholia.

Mourning is regularly the reaction to the loss of a beloved person, or to the loss of some abstraction which has taken place of one, such as one's country, liberty, an ideal, and so on. In some people the same influences produce melancholia, and we consequently suspect them of a pathological disposition.
Mourning and melancholia 1917, trans. J. Strachey in vol 14 of Freud *Works* Hogarth Press 1953–73

He summarizes the features of melancholia:

a profoundly painful dejection, cessation of interest in the outside world, loss of the capacity to love, inhibition of all activity, and a lowering of the self-regarding feelings to a degree that finds utterance in self-reproaches and self-revilings, and culminates in delusional expectations of punishment.

The process of mourning, according to Freud, involves the slow detachment of libido from the loved object:

Each single one of the memories and expectations in which the libido is bound to the object is brought up and cathected, and detachment of the libido is accomplished in respect of it. . . When the work of mourning is completed, the ego becomes free and uninhibited again.

This process presumably underlies mourning rituals: to allow the mourner to experience loss intensely so that he may finally return to the world of the living again and function effectively in it. Melancholia differs from mourning in being more diffuse and intractable. The melancholic doesn't quite know what he/she has lost. Suicidal thoughts may appear, destructive impulses towards an ambivalent figure, loved and hated but now lost, so that the feelings have nowhere to go but against the self. Hamlet is a classic case of melancholy, and was in 17th-century medicine too. Samson is a less clear example. Who or what has he lost? His country? Liberty? God? Dalila? *Damon* and *Samson* have important affinities, with roots in the same dark, incomprehensible experience: M's dark night of the soul.

Lines 1–17 Introduction
The introduction expresses strong feelings of grief but is tightly controlled. The essence of the psychological narrative is compressed into these 17 lines. The poem is first located in time and place. *Himerides*, the first word, establishes a link with Italy and the pastoral tradition. This leads directly to mention of Bion, pastoral elegist and subject of pastoral elegy, a fusion of M and Diodati of more than 1000 years ago. *diu*, long, is ambiguously placed, to refer to both their memory (which lasts till the present) and the tears, which lasted a long time but had an end. The pastoral tradition includes a society of living and dead, which guarantees a kind of immortality to the dead who are included in it. Other kinds of time are also important: the rhythm of the seasons (a constant in the world of Bion as of M), human life (Damon's death was premature), and ritual. Thyrsis was absent at the moment of death, and two years have passed without formal mourning. The rituals of mourning have been defective. The rest of the poem has to complete the mourning-work, to allow M's ego to become 'free and uninhibited again'.

Two other oppositions are touched on in the juxtaposition *Sicelicum Thamesina*, Sicily and the Thames, Italy and England. England was another reality M had to face up to. It is Thames' *towns*, too, which must hear this song, an affirmation of city against country, urban values against pastoral, foreshadowing the rejection of pastoral at the end of the poem.

35

On M and the pastoral tradition, read the section on pastoral in *Lycidas*, in *Odes, pastorals, masques* in this series. Especially relevant for *Damon* are: Theocritus *Idyls* I and VII, Bion *Idyl* I, Moschus *Idyl* III (Lament for Bion), and, most important, Virgil *Eclogues*.

19–34 Celebration of Damon
The first part of the ritual seems easy to perform, Damon's transition to the world of the dead and his incorporation there. M prays for his welcome in heaven (though he doesn't imagine it yet) and undertakes to look after his fame on earth, so Damon's new relationship to living and dead seems fixed and assured.

45–67 Separation
M then turns to himself, experiencing his own loss first through recreating memories of the past as intensely as he can, then describing his demoralized state. He wanders alone, separated from society, uninterested in the external world – though the images of lines 58–61 are brilliantly observed, with a realism that stands out in the pastoral context. Or does it?

69–92 Failed transition
In this state he is approached by three types of comforter who seek to re-incorporate him into society before he is ready. *Samson* has the same three types, male friends, elder male, and attractive female, and it is interesting to compare them. Here the friends, Tityrus etc, are seductive male friends (especially Amyntas). Samson's friends, however, are undifferentiated, 'certain friends and equals' to whom Samson reacts with indifference. The friends in *Damon* are followed by Mopsa, a pedant, physician, moralizer, a Polonius-figure rejected with swift contempt. (Manoa, Samson's father, is treated with careful respect by Samson.) Then follow the nymphs, whom M can't allow to take the place of his beloved. Some of the details of 88–92 sound so specific that these were probably actual girls pushed in his direction by his father or concerned relatives. They seem the antithesis of Dalila, since they are recommended by his society, rejected by the individual: but M's response is not unlike Samson's.

94–160 Transition
As in *Samson*, this confrontation ends in rejection but seems to have a cathartic effect. M begins a curious piece of reasoning. Man is contrasted with animals on two contradictory grounds.

Animals form pairs or groups more readily, ie they are more social, but they also form new relationships more readily, ie the particular bonds are weaker. M is drawing a lesson from this, or asserting an ideal, of a confused kind, of which the upshot is that men must be as callous as animals. There must be a limit to mourning. It is interesting that a similar speculation occurs in *Samson* 1046–52, after Dalila has gone. Here the rare gift of heaven is explicitly a wife, as M's metaphors suggest is the case with *Damon*, though the friend in question is male.

Lines 125–138 have seemed offensive and irrelevant to some critics: an ego trip. That is precisely what they are, in a strict sense. The guilt of his absence and consequently defective ritual need to be expiated. Why did I go? he asks. But he did go, and the journey was deeply satisfying, an emotional fact that he needs to admit to himself. The crucial separation took place *before* Damon died. Mention of this separation allows him at last to imagine Damon actually dead, himself performing the ritual act of closing the eyes of his friend. After this his normally strong ego can reassert itself, as he acknowledges how much he valued the Italian experience. This is followed by his last re-creation of his friendship with Damon. This is different in kind from all previous ones, because here it's clearly recognized as unreal, occurring after his death.

The poem now approaches the turning point. Round line 155 language breaks down, regrouping round the fixed point of a re-formed ego. *Ipse* (myself) is symbolically detached from the dead Damon and attached to M, and repeated again at line 162, *Ipse ego*, M now proclaiming his ability to write poetry again and participate in the social world. In Freud's words, 'the work of mourning is completed, the ego becomes free and uninhibited again'.

162-end Incorporation

M's announcement of his epic vocation is in fact a rededication, since he had already made this general commitment in *Elegy VI*, written to Diodati. What is new is that it is to be a *British* epic. This signifies his reconciliation to England, his willingness to be recincorporated into the land of his birth.

The cups/books Manso gave him then become the pretext for a vision, in which Damon is fully incorporated into the world of the dead. Manso was an Italian nobleman, patron of poets, a father-figure. (In his poem to him, M calls him 'father Manso'

Mansus 25.) M has the blessing of this figure, as Samson did of his real father, Manoa.

The vision when it comes is startling. Some critics have been shocked by the sheer release of libidinal energies in it. Damon's virginal virtue is to be rewarded with the 'honours due to a virgin', but what would the honours due to a virgin be in this wild Bacchanal? Damon has been absorbed into Dionysos, in an existence ruled by the Dionysiac *thyrsus*, not the 'golden rod' of judgment of line 23. Damon is now separate from M, but M has also formally renounced the Dionysiac principle in himself along with Diodati. It has been pointed out that this vision is not particularly Christian. Heavenly love is Christ, sublimated Eros, and some of the details come from the book of *Revelation*, but it takes more than this and the word *Sionaeo* to neutralize the energies of the poem's final image. *Samson* and *Damon* both end in a wild feast. Samson destroys it, Damon joins it: two kinds of oblivion.

English sonnets: Sonnets 7–8

> In his hands
> The thing became a trumpet, whence he blew
> Soul-animating strains; alas! too few.

Wordsworth's high opinion of M's sonnets was not shared by Dr Johnson 'Of the best of them the most that can be said, is that it is not bad.' The sonnet-form was a surprising one for M to be attracted to. Not just because the fashion for sonnets had been over for 30 years – M never worried about fashion – but because it is so strict a form, and so small in scale. Perhaps this is what appealed to him. Most of them were written between his two major periods of poetic activity, in the decades 1640–60, when he was immersed in public life, and had lost his way poetically. The sonnet-form allowed him to explore personal doubts and anxieties within a well-defined and demanding form, its constraints accepted and mastered being a guarantee of M's own over-all control. It also allowed him to intervene briefly and directly on public themes, relying on a massive simplicity of diction and thought.

Outside M some very great poetry has been written in the sonnet-form: worth reading for its own sake, as well as for any light it may throw on M's use of the form. The sonnets of

Cavalcante (died 1300), Dante (1265–1321) and Petrarch (1304–74), which created the vogue for sonnets that at one time dominated the European literary scene, still have the power and immediacy they first had, now that the masses of their imitators have disappeared from view. Della Casa (1503–56) was particularly influential on M. In England, Wyatt's sonnets are the first major example of the form, fine poems with a complexity of syntax and rhythm that is good preparation for reading M's. Shakespeare's sonnets were certainly known to M, though they seemed not to have impinged as a model or challenge.

As it became conventionalized, the sonnet form acquired strict rules, the 14 lines being divided into an octave (8 lines, usually grouped 4–4) and a sestet (6 lines, either 3–3, or 4–2). There was more variation in the sestet, the rhyme patterns tending to either a closed form, with a concluding couplet to wrap up the poem, or a more interwoven form. M characteristically used the basic structure with considerable freedom, his complex syntax often in tension with the tidy divisions of the form. Here is Wordsworth on M's use of the form:

Milton, however, has not submitted to this [ie the pattern of octave and sestet]. In the better half of his sonnets the sense does not close with the rhyme at the eighth line, but overflows into the second portion of the metre. Now it has struck me, that this is not done merely to gratify the ear by variety and freedom of sound, but also to aid in giving that pervading sense of unity in which the excellence of the sonnet has always seemed to me to consist. Instead of looking at this composition as a piece of architecture, making a whole out of three parts, I have been much in the habit of preferring the image of an orbicular body, a sphere, or a dew-drop.   Letter to Dyce 1833

Sonnet 7 How soon hath time
M included this poem in a much-revised letter to a friend, probably his former tutor, the stern Calvinist, Thomas Young:

You are often to me, and were yesterday especially, as a good watchman to admonish that the hours of the night pass on (for so I call my life as yet obscure, and unserviceable to mankind) and that the day with me is at hand wherein Christ commands all to labour while there is light.

The poem is offered as an early fruit, proof that M had already thought of his friend's admonitions. But the poem throughout is obscure, private, excluding the reader. The first four lines pose the problem. M is getting old and has nothing to show for it. The next quatrain, however, is shrouded in qualifications, uncertain in meaning. The resolution then is a masterpiece of stately, self-confident obscurantism. The rhythm moves with a firm steady tread, like the inexorable progression of time itself, accommodating all extremes of space and time. But what is 'it' of line 9? No-one is sure: probably not even M. Some have suggested 'ripeness' but can ripeness be 'more or less'? Can you say 'soon ripeness' or 'slow ripeness?' By a trick of grammar M has resolved his problem by handing over to God the task of defining what 'it' is. 'Lot' similarly defeats grammar. It's both a quantity (of fruits?) and a place towards which Time (no longer a thief) is leading M.

The final couplet has similar difficulties. 'All' has the same vagueness as 'it'. 'Is' is in the present tense, the universal present, God's time which is always present, but this is immediately contradicted by the if-clause that follows: certainty cancelled by doubts. This is the paradox of Calvinism. God has predestined everyone to be saved or damned from the beginning of time, but no-one knows what His inscrutable will holds for them. So M is certain, but also knows that he cannot be certain. The real question is not whether he is doing reasonably well at this stage, but what God has decided about him from the beginning of time.

Sonnet 8 Captain or colonel

This poem was probably written in Nov 1642, when the royalist army approached London but was turned back by the parliamentary forces led by the Earl of Essex. At this point in the war, the king's defeat was by no means certain, and the parliamentary leadership was still predominately aristocratic. M's tone is curiously detached, though, in keeping with the poem's theme. It starts in the present – captain and colonel were current military ranks – but 'knight in arms' begins the slide into the past and/or literature. M in 1642 was a supporter of parliament, but is still an elitist, addressing himself to officers only, assuming in them a high level of culture, so that they will preserve a poet against the barbaric impulses of their own troops: a natural alliance of members of the cultivated classes against the brutish lower orders. The poet himself as a self-interested individual has almost disappeared from the poem. He asks for protection for the doors first, and the poet is 'he' not 'me'. The poets he refers to, Pindar and Euripides, were both dead at the time of the incidents described. It's as though M was talking about himself posthumously.

A kind of love? Sonnets 9–10

Sonnet 9 Lady, that in the prime

Who is the 'Lady' of this sonnet? One name that has been suggested is Mary Powell, who married M probably in the mid-year of 1642 after a very brief courtship, and returned home to her parents after a few weeks. M considered divorcing her, but she returned to him in 1646. M was 33 at the time of his marriage, learned, pious, repressed, totally naive about women. He certainly misjudged young Mary's character, as he later admitted bitterly. If this sonnet was written for her, it becomes a clumsy kind of love poem, the only one he wrote to her. This would give its self-righteous, moralizing tone a pathetic quality, an interesting poem for being so wrong about the lady (and himself, since the qualities he praises in her are ones he valued in himself).

Some details gain point if this is Mary. 'Mary' is mentioned in line 5. This is the Mary who chose to listen to Christ while her sister Martha did the housework (*Luke* x). She was normally taken to represent the virtues of contemplation. Addressed to Mary Powell, wife-to-be or new-married bride, this would suggest that she has chosen to listen to her husband's delightful conversation rather than do the housework. This would be ironic in the light of M's disappointment in Mary's intellectual interests and capacities. The reference to Ruth makes Mary Powell seem even more likely as the subject of the poem. Ruth's supreme virtue was that she followed Naomi, mother of her dead husband, back to Israel, showing her total commitment to the family of her husband, not her own. Ruth was rewarded by becoming the ancestress of David. Mary Powell, however, travelled in the opposite direction to Ruth. Was this poem an inappropriate compliment? Or a plea for her return?

The conclusion envisages her death, but gives it a muted erotic quality similar to *Damon*, though less intense. The bridegroom, Christ, 'passes to bliss' (which sounds stronger than a feast), and she has 'gained [her] entrance'. Entrance to what? If this was Mary, the eroticism seems strangely muffled by biblical allusions, but if it was not it becomes a more curious performance. Unfortunately there is no evidence that could help us decide whether the lady was Mary or someone else.

Sonnet 10 Daughter to that good earl
There may be an undercurrent of erotic feeling below the surface
of this poem, too. Lady Margaret Ley, daughter of the Earl of
Marlborough, had married Captain Hobson, a parliament supporter, in 1641. M's nephew, Edward Phillips, describes M's
relation with her over this period (1642–5): 'Our author now
as it were a single man again, made it his chief diversion now
and then in the evening, to visit the Lady Margaret Lee, daughter
to the Earl of Marlborough.' The relationship was very proper,
and included the husband. But in the poem there's no mention
of Captain Hobson, just the aristocratic and dead father, who
takes over as the real subject, into whom the daughter is transformed: totally safe, an object of honour not of love at all.

In fact 'that good old earl' did not seem good to everyone
who knew him. 'An old dissembler, wont to be called Volpone,
and I think he as well deserveth it now as ever', wrote one
contemporary. Certainly he was an aristocrat and servant of
kings. Was M rethinking his allegiance to parliament over this
time?

The problems of liberty: Sonnets *11*, *12* and *New forcers*

These three poems, written between 1644 and 1647, come out
of M's growing dissatisfaction with the direction the civil war
was taking, under the direction of the dominant presbyterian
faction in parliament. M had initially supposed that his side was
fighting for 'liberty' against all forms of tyranny. But his
separation from Mary Powell hastened on his political education.
His bitter experience set him thinking and writing about alter-
natives to prevailing marriage customs. But his ideas were
greeted with hostility by the presbyterians, and M realized that
the 'freedom of the conscience' they had seemed to stand for
was merely freedom for everyone else to think what they thought.
They were now as ready to use state power to back their ideology
as the hated Archbishop Laud had been, whom they had had
executed.

But M had another problem about his divorce tracts. The
people who agreed with them were as bad as those who opposed
them. The contradiction can be seen in the two poems *11* and
12. *Sonnet 11* is directed against presbyterians who are so pro-

vincial and ignorant that they do not read beyond the title, and reject *Tetrachordon* unread on false grounds of style. *Sonnet 12* has similar but greater contempt for the stupidity of its readers, but this time it's because they *have* read his works. M's outburst is directed against people like Mrs Attaway, who left her husband and praised M's writings on divorce for justifying the idea. Line 10 contains a typical piece of Miltonic thinking: 'Licence they mean when they cry liberty'. This sounds a fine moral distinction, but in practice it rests on an authoritarian system, since it has to be administered by the 'wise and good'. Who are they? Who decides? M is caught here in the same contradiction as he saw in the presbyterians.

On the new forcers is the weightiest and most famous of this group. It is less self-concerned and, instead of a snobbish display of erudition, M has deployed his studies in language and history to carry his central political insight, brilliantly concentrated in the final sting-in-the-tail. Formally, M has departed from the traditional sonnet structure, drawing on a variation used by Italian satirists, the 'sonetto caudato', the tailed sonnet. In the history of the word 'priest', its corruption from the Greek 'presbyter' or elder and its current replacement by its original form, M sees the history of the presbyterian faction, the transformation of a successful rebellon into a new oppression. Orwell made the same point about the Russian revolution in *Animal farm*, but M puts it with the indignation of a revolutionary who still believes in the validity of his revolution.

Unpolitical sonnets: 13, 14

Sonnet 13, to Henry Lawes, was written in 1645 and published in 1648 with Lawes' *Choice psalms*. This makes it a political act, since Lawes dedicated this book to King Charles, then in prison, shortly to be executed and then denounced by M on behalf of parliament. M's motives no doubt included personal and artistic respect for Lawes, an admirable composer who had written settings for M's own *Comus* and *Arcades*, but he showed political courage in making that respect public at this time: proof of how sincerely he believed that art transcended political faction.

Sonnet 14, on Catherine Thomason, was another politically neutral poem from this period. George Thomason her husband, was a royalist and lover of learning, a bookseller by trade. It has the classic form for a poem on death, separation (1–4), transition (5–11), incorporation (12–14), but it lacks the double concern of a death-ritual, with the living as well as the dead. Thomason is not mentioned, so he has no place in the ritual. M makes it feel even more remote by putting it all into the past tense. Even the day of judgment has already occurred for her, in lines 13–14. Like the fair infant, she has been removed from human history.

In search of a saviour

Sonnets to Fairfax, Cromwell, Vane
These sonnets map M's political development between 1648 and 1652. In each case M looks desperately to a single figure to save England and the revolution, an Apollo or Samson. The choice of figures shows M's drift to the left. Fairfax was the leader of the New Model Army which defeated the king's forces, but Fairfax was a moderate who disapproved of the execution of the king. Cromwell supported the execution, but afterwards put a brake on the revolution, resisting the more extreme demands of the radical independents. Sir Henry Vane the younger's political position was less clear, but he was identified more closely than Cromwell with the independents, and was to be executed after the Restoration.

Their form also has common features. They all start with a a single name, the surname, with no titles or indications of affection or respect. Contact is established, but not a relationship, personal or public. The name acts as a fixed point round which the poem accumulates over the first 8 lines. In the poems to Fairfax and Cromwell, the last 6 lines urge a programme which in effect repudiates the grounds of their previous achievement. Fairfax's name 'in arms through Europe rings', but M wants him now to repudiate war and turn his attenton to England. So also with Cromwell, urged to win the victories of 'peace' – though this is less of a contradiction, since 'peace' really refers to a holy war waged against the militant presbyterians. What has happened to the M who condemned those who 'adjure the civil sword To force our consciences'? One thing is that he hopes he can rely on Cromwell's support. But M can't believe that his opponents have consciences to force in the first place. They are simply wolves, to be exterminated.

The *Sonnet to Vane* confronts his contradiction directly. The problem is resolved by the claim that Vane's solution to it is definitive, though M doesn't say what that solution is. Vane resolves a number of oppositions, youth and age, diplomatic and military strategy, peace and war, and finally spiritual and civil power. The claim that 'the bounds of either sword to thee we owe' is either wishful thinking (Cromwell not Vane had his hand firmly on the swords concerned) or it refers only to a theoretical clarification. Thanks to Vane we know where these bounds lie: now it is up to us (or Cromwell) to put this knowledge into practice. The first alternative would be subversive, and there are other such hints in the poem. The superiority of gowns to arms suggests the superiority of the diplomat (Vane?) to the soldier (Cromwell?). 'Eldest son' could see Vane as an heir apparent, ready now to take over political power. But though M might hint at this idea in a poem for Vane's own eyes, he would not have dreamed of stating it publicly and unequivocally in 1652.

Further reading. Marvell's *Horatian ode* and *Appleton House: to my Lord Fairfax*, subtle and profound explorations of the significance of Cromwell and Fairfax, are major poems to be read alongside M's two sonnets.

Sonnet 15 On the late massacre in Piedmont
This is another political poem, but its impact is more direct and focused than the three previous sonnets. Those poems tried to resolve the problems of a very complex situation by concentrating on a single figure whose personal resolution of oppositions might transform public life. This sonnet however is addressed directly to God (though M might have hoped that Cromwell would overhear), and is concentrated by a single unequivocally horrendous incident, a 17th-century equivalent of the massacre at Mai Lai in the Vietnamese war. On 24 April 1655 the Duke of Savoy in northern Italy had sent troops against the Waldensians, a religious sect in Piedmont whose doctrine seemed protestant in form, but derived from a medieval founder, Peter Valdes. Protestant Europe was outraged, and Cromwell took up their cause. Milton, as his secretary for languages, drafted letters of protest sent to the Duke of Savoy and other European heads of state.

The power of the poem comes from the simplicity of the basic feeling of outrage, unqualified by any moral complexities. The Waldensians are completely innocent and good, the Piedmontese soldiers wholly evil. The address to God rather than

45

Cromwell also allows feelings to be expressed more directly. M uses imperatives to God: he can only suggest to Cromwell. The price of this directness is that he has a prayer but not a policy.

But the poem is not a self-indulgent rush of emotion. 'Avenge O Lord thy slaughtered saints', M begins, then goes through these elements in reverse order, describing the results of the slaughter, establishing that they were saints, then restating 'avenge' as 'forget not'. This is the moral progression of the poem. M's indignation isn't expressed as a desire for bloody retribution. The particulars of the outrage are not blurred but are placed in an over-all cosmic scheme in which the final triumph of the good is assured. So the lesson the poem ends with is not how to wage guerilla warfare on the Alps, but advice to 'fly the Babylonian woe'. M's faith was vindicated militarily on this occasion, for the Waldensians won a decisive victory in July of that year. M's response is worth comparing with what would be possible today, if you were to write on a similar atrocity. Can we in the 20c believe in a God who will carry out the necessary retribution? Can we believe that history will vindicate the deaths of the innocent, civilian casualties in North and South Vietnam, protestants and catholics in Northern Ireland? It's also interesting to compare the mood of this poem with *Samson*, which ends in cataclysm. Total destruction of the enemy is the satisfaction Samson demands of God, and receives.

Sonnets against despair: 16–19

M's last five sonnets form a natural group, personal and domestic in concern, intimate in tone, concerned with responses to loss, coming to terms with blindness, age and loneliness.

Sonnet 16 When I consider is the earliest. Its date has been the subject of scholarly dispute. The most favoured dates are: 1645, 1651/2, or 1655. The decision matters, because it involves decisions about the central theme of the poem. Take the key phrase 'my light is spent'. What does 'light' refer to? Eyesight? So we can ask when M was most troubled about his blindness. His eyes were weak from youth, but 1644/5 marked a decisive stage. In a letter to a man named Philaris in 1654 he recalled 'It is ten years more or less I think since I felt my sight getting weak and dull, and at the same time my viscera generally out of sorts.' The general illness he felt may have come from the punitive cures that he had to endure. In 17th-century medicine the cure was often worse than the disease, and his final blindness was at least a relief from painful treatment. Total blindness came in late 1651/early 1652, hence the common wish to date the poem then or shortly after.

But consider other things the phrase could refer to. Sexual powers? M was without a wife in 1644, 1652 and 1655. Intellectual powers? In 1644 M was going through a crisis of allegiance. In 1652 he had published a major work, but lost his eyes doing so. Poetic powers? Between *Damon* of 1640 and *PL*, begun some time in the 1650s, M's poetic output dried up (though the date of *Samson* isn't known).

Behind the poem are the same Calvinist anxieties, and even some of the same Biblical texts, as behind *Sonnet 7*. This poem reads like a bleak interim report, 12 or 20 years on. The reference to 'that one talent' is ominous. In the parable, the man with only one talent buried it, and lost even that. 'From him that hath not shall be taken away even that which he hath' (*Matt* xxv). This is the demanding God of *Sonnet 7*, the 'great task-master'.

M's duty, as one of God's elect, was to illuminate the sinful world. 'Let your light so shine before men that they may see your good works, and glorify your father which is in heaven' (*Matt* v). The light that is 'spent', then, contains a cruel ambiguity: the good works done by M, and his eyesight, quenched by God. One contradicts the other. M can't be a light to the world, because he can't see. So God has given him a command, then taken from him the means of carrying it out. M confronts the blasphemously rebellious thoughts that this rouses in him, and the rest of the sonnet tries to allay them.

Part of M's solution is to re-present the problem in terms of a different economic and social system. The God of the first half is a grasping capitalist, interested only in maximizing his profits. Men are his business managers who must speculate wisely and accumulate interest on God's capital. This image is seemingly forced on M by the words of the Bible. However, the God of the second part is not a capitalist but a king ('his state Is kingly') to whom men are bound by something more like feudal bonds. They must pay him by service, not by money, and contribute to his state, to his capacity to make a conspicuous display. The service demanded is decorative rather than useful. So M's religious anxieties, it seems, could be allayed only by imagining the politics of heaven cast in the form of the reactionary regime he had devoted his political energies to overthrowing. It would be very interesting to know if he wrote the poem in 1644–5, or after 1649, when he was publicly committed to the execution of the king and the deposition of the Stuarts.

None of these considerations is decisive about the date. The majority of scholars favour either 1651–2 or 1655, but the most direct evidence is the words of the poem, 'ere half my days'. If M had the Biblical three score and ten (70) in mind, this places it at 1543 or before.

Sonnet 17 Lawrence

The date of this poem is uncertain, too, but nothing hangs on the difference. The important question is the nature of the relationship. Young Edward Lawrence was 25 years younger than the middle-aged poet: 18 years old in 1651, the earliest plausible date, and 24 when he died in 1657. M addressed him formally, using his surname only, and refers to his and his father's virtue, reminiscent of the public sonnets *15–17*, and the careful formality of the *Sonnet to Lady Margaret Ley*. But the formula 'of virtuous father virtuous son' has an interesting classical source. Horace had written a verse epistle to a young girl beginning

> O maiden, fairer than your mother fair,
> make any end you will of my
> abusive lines, be it with fire
> or in the Adriatic's waters.   *Odes* i 16

an old man wryly propositioning a pretty girl, pre-empting the expected refusal.

The rest of M's poem is a tentative attempt to recreate the kind of relationship he once had with Diodati: Charles' role had been to urge the serious young M to enjoy life. M gives the same advice to young Lawrence, in a touchingly ponderous and slightly ambiguous form.

Sonnet 18 Cyriack, whose grandsire gives similar advice, but the tone is very different. Cyriack was a pupil of M's, later almost a disciple. M confidently uses his first name, and the poem is full of imperatives: 'Today...resolve' etc. To Lawrence, M used only tentative questions, not even committing himself or Lawrence to a definite invitation.

M addressed another sonnet to Cyriack, who didn't seem to get M's best poems written for him. This one, on M's blindness, has none of the power of honesty of *Sonnet 16*. He claims now that he 'argue[s] not Against heaven's hand or will'. If he did, he wouldn't tell young Cyriack about it.

M's most powerful treatments of blindness elsewhere are the invocation to light at *PL* III 1–55 and *Samson* 65–109.

Sonnet 19 Methought I saw

Again there is a problem of dates. Which 'saint' is this? Katherine is the more likely, whom M married in 1656. She died in 1658, after giving birth to a son. His first wife Mary had died in 1652, having given birth to a daughter. In favour of Katherine are a number of apparent allusions, to her purity (her name comes from a Greek word meaning pure) and to his vagueness of perception (he had never seen Katherine.) It's also assumed that M could not have loved Mary Powell, so he must have loved Katherine. This last is a dubious assumption. Mary has had a bad press from Milton scholars, but his marriage to her did last 10 years after its shaky start, and at least he had seen what she looked like.

The poem ostensibly starts from a dream. Whether or not there was a real dream-experience behind its composition, M could draw on a long tradition of dream-poems. Does that make it less sincere? Dreams are private, opaque, and usually transient, dissolving on confrontation with the waking world. M may have needed this framework to give his dream the needed permanence and public form it needed.

The basic structure comes from Euripides' *Alcestis*. Death came to claim Admetus, but was persuaded by Apollo to spare him if someone would die in his place. Everyone refused to, including Admetus' old father and mother, who clung to what remained of their life. Only his wife Alcestis was ready to die instead of him. But on the day of her death, Hercules, a friend of Admetus, passes by. In spite of his grief Admetus feasts him. Hercules finds out what has happened, goes to the tomb and wrestles victoriously with Death, and brings Alcestis back to her husband.

M's first quatrain retells part of the Alcestis story. The second quatrain seems to leave Euripides for the OT, though the Greeks had purification rites too. But the point of this reference is not clear. He seems to be expressing a disgust at the pollution of child-birth (spot, taint), but he uses a syntax so tangled that no further logical point emerges. Is the husband of a woman who dies as a result of giving birth somehow implicated in the death? Admetus was guilty of Alcestis' death in Euripides' play.

49

The vision then intensifies. She almost appears before him, her face 'veiled'. Is this because the face is Katherine's, which he'd never seen? Alcestis also had been veiled. Hercules had brought her back, to test Admetus, who had sworn never to marry again. Hercules persuaded Admetus to take this 'slave' into his house: to be unfaithful to his wife with his wife. The test is highly ambiguous. Admetus is persuaded to continue with living, simultaneously breaking faith with his previous wife and remarrying her. In the real world, where dead wives remain dead, Euripides' play could suggest the lesson that to re-marry is a higher kind of faithfulness. M did remarry, after the deaths of both Mary and Katherine.

It's important to remember that although M probably wrote the poem at a particular date with a particular wife in mind, dreams work by a different kind of logic. Two or more figures characteristically merge into a single dream-figure, who has qualities of both. So if M had Katherine uppermost in his mind, as seems more likely the wife-figure could still be filled out with memories of Mary. Perhaps other women too. Broadbent notes how the poem refers to her in 'terms which are profoundly maternal' (*Calm of mind* ed J. R. Wittreich, Cleveland, 1971). He points out the importance of sons. 'Jove's great son' is the rescuer, a son more powerful than the husband. Admetus, remember, wished his parents to die for him, and they refused. Now the 'saint' has died giving birth, but is purified of birth-pollution (hence no longer a mother but a wife?) and is almost available to M. Almost, but not quite, and she flees.

The language of the poem is unusually simple for M, very monosyllabic (87 per cent monosyllables, with only one poly-syllable per line, compared to 60 per cent monosyllables, 2 polysyllables per line for the previous sonnet, *To Cyriack Skinner* 22). He'd remembered the English language at last, though the syntax is difficult, and the bare words reverberate against the complexities of classical myth. The language is also imprecise, not only visually (shapes and qualities blurred as a blind man might see them) but also emotionally. He doesn't actually say he loves or loved whoever it was. Once he says he 'trusts', but that emotion is directed towards God, and there is one 'O'. Otherwise the weight of his feeling comes over without him talking about it at all. It remains a reserved, impenetrably private poem.

The Shorter Poems

Elegia I ad Carolum Diodatium

TANDEM, CARE, tuae mihi pervenere tabellae,
    Pertulit et voces nuntia charta tuas,
    Pertulit occidua Devae Cestrensis ab ora
Vergivium prono qua petit amne salum.
Multum crede iuvat terras aluisse remotas   5
    Pectus amans nostri, tamque fidele caput,
Quodque mihi lepidum tellus longinqua sodalem
    Debet, at unde brevi reddere iussa velit.

Me tenet urbs reflua quam Thamesis alluit unda,
    Meque nec invitum patria dulcis habet.   10
Iam nec arundiferum mihi cura revisere Camum,
    Nec dudum vetiti me laris angit amor.
Nuda nec arva placent, umbrasque negantia molles,
    Quam male Phoebicolis convenit ille locus!
Nec duri libet usque minas perferre magistri   15
    Caeteraque ingenio non subeunda meo.
Si sit hoc exilium patrios adiisse penates,
    Et vacuum curis otia grata sequi,
Non ego vel profugi nomen, sortemve recuso,
    Laetus et exilii conditione fruor.   20
O utinam vates nunquam graviora tulisset
    Ille Tomitano flebilis exul agro,
Non tunc Ionio quicquam cessisset Homero
    Neve foret victo laus tibi prima Maro.

1 **tandem** (at last) this line is direct in expression, contrasting with the jokey elaboration of the next line, which repeats its idea. 2 **voces** (words or voice) living speech, contrasted with *charta*, paper, which is dead, but acts like a living person through *nuntia*, announcing, messenger-like. 3 **Devae** (river Dee) called 'wizard stream' in *Lyc* 55. Diodati is living near Chester, having recently graduated from Oxford. 7 **tellus** (the Earth) used for a district of Britain, suggests that it's like another world. 9 **reflua** (flowing back, ebbing) Thames flowing backwards, and in Diodati's direction, in contrast with the Dee. 10 **patria** (paternal) his father adjectivalized, and made feminine in gender, agreeing with *domus* (home) understood. 12 **vetiti** (forbidden) see Commentary p. 12. Sounds too strong to refer to absence during vacation. **angit Amor**

Elegy I   To Charles Diodati

AT LAST, DEAR CHARLES! a letter from you has arrived,
eloquent sheets of paper transported your words,
brought from the western bank of the Dee, near
Chester,
where it seeks the Irish sea with an urgent stream.
It truly delights me that lands so remote have nourished   5
a heart so loving towards me, a mind so true,
that the distant world that owes me an elegant friend
will soon return him willingly, on demand.

I'm still in the city washed by the ebbing Thames,
gladly detained in my sweet paternal home;   10
Not anxious yet to return to the reed-rich Cam,
nor choked with love for rooms they've just forbidden me.
For naked fields that lack soft shades are unpleasing –
that place is ill equipped for Apollo's shrine:
And the endless threats of my harsh tutor are irksome,   15
and other inflictions my nature will not endure.
If exile is this, to come to my father's house,
free to pursue a life of pleasure and ease,
I formally welcome the title and fate of an outcast,
happy to make good use of my exiled state.   20
I wish that Ovid had suffered no worse in the time
he wept in exile, tending a farm in Tomis:
he'd not have yielded the palm to Ionian Homer,
but surpassed you, Virgil, and won the highest praise.

(love grips) sarcastic exaggeration. **13 negantia** (refusing) Cambridge
isn't treeless, so this must refer to a mental landscape, lacking shadows,
mystery, imagination. **15 magistri** (master, teacher) Chappell. See Commentary p. 12. **20 fruor** (take pleasure or profit from) the word combines
both sides of normal puritan opposition between pleasure and usefulness.
21 vates (bard) Ovid, banished by Augustus to Tomis, partly for the
immorality of his verse. *vates* unites notions of prophet and poet,
suitable for M's religious sense of his poetic vocation, but not Ovid's.
24 Maro (Virgil) the idea of epic poets yielding the palm of victory to
erotic poets and poetry was commonplace in Latin love poets; eg
Propertius I 7 & 9.

Tempora nam licet hic placidis dare libera 25
    Musis,
Et totum rapiunt me mea vita libri.
Excipit hinc fessum sinuosi pompa theatri,
    Et vocat ad plausus garrula scena suos.
Seu catus auditur senior, seu prodigus haeres,
    Seu procus, aut posita casside miles adest, 30
Sive decennali foecundus lite patronus
    Detonat inculto barbara verba foro,
Saepe vafer gnato succurrit servus amanti,
    Et nasum rigidi fallit ubique patris;
Saepe novos illic virgo mirata calores 35
    Quid sit amor nescit, dum quoque nescit, amat.

Sive cruentatum furiosa Tragoedia sceptrum
    Quassat, et effusis crinibus ora rotat,
Et dolet, et specto, iuvat et spectasse dolendo,
    Interdum et lacrymis dulcis amaror inest: 40
Seu puer infelix indelibata reliquit
    Gaudia, et abrupto flendus amore cadit,
Seu ferus e tenebris iterat Styga criminis ultor
    Conscia funereo pectora torre movens,
Seu maeret Pelopeia domus, seu nobilis Ili, 45
    Aut luit incestos aula Creontis avos.

Sed neque sub tecto semper nec in urbe latemus,
    Irrita nec nobis tempora veris eunt.
Nos quoque lucus habet vicina consitus ulmo
    Atque suburbani nobilis umbra loci. 50
Saepius hic blandas spirantia sidera flammas
    Virgineos videas praeteriisse choros.

26 mea vita (my life) usually a term of endearment, which here could be addressed to his books, or more probably to Diodati. **27 sinuosi** (winding, or billowing, swelling) compressed construction. Theatre is winding, intricate, so is the procession, and billowing with inflated language. **29** stock characters from Roman comedy, which M would read rather than see, though there were performances at banned Cambridge. **43 ultor** (revenger) the ghost of Hamlet's father? Seneca's

For here I have time to give to the gentle Muses,   25
    and books my darling possess my very soul.
Till I tire, and a swelling theatre throng entices me,
    the chattering stage requires that I come and applaud.
The speaker's a cunning old man, or a spendthrift heir;
    or a suitor appears, or soldier with helmet undone;   30
or a lawyer, fat with the fruits of a ten-year suit,
    thunders his barbarous terms at an ignorant court:
And often a sly slave furthers a son's romance,
    tricking the strict father at every turn;
and many a maiden, amazed by strange new feelings,   35
    knows nothing of love, but loves in her ignorance.

Or Tragedy shakes her gory sceptre, and rages;
    her hair dishevelled, she stares wildly around.
There is pain but I look, there is pleasure in looking in pain,
    at times the taste of tears is bitter-sweet.   40
We weep when a star-crossed youth must leave his
        delights
    untouched, and dies, his love torn from his side:
or a fierce avenger of crimes returns from the shades,
    arousing guilty hearts with brands from his pyre;
or the grief of the royal families of Greece and Troy,   45
    or the harsh atonement for incest of Creon's court.

But I'm not for ever hiding indoors, in town;
    for me the pleasures of spring don't come in vain.
And a nearby glade thick-planted with elms receives me,
    and places famous for shade close to town.   50
From here you often can see, like stars that exhale
    soft flames, parties of girls go dancing by.

Thystes, the Roman ancestor of Elizabethan revenge tragedy, sounds
more likely. **45 Pelopeia** (Pelops') ancestor of Agamemnon, leader
of the Greeks against Troy, subject of Aeschylus' *Oresteia*. **Ili**
(Ilius') a king of Troy, a less fertile source for tragedies. **46 Creontis**
(Creon's) ruler of Thebes, character in Sophocles' *Oedipus* series. **50
suburbani** (near the city) plural in form but could refer to a single
specific place. See also *El VII* 52. Perhaps a property belonging to
a wealthy friend of his father's.

Ah quoties dignae stupui miracula formae
  Quae possit senium vel reparare Iovis;
Ah quoties vidi superantia lumina gemmas,   55
  Atque faces quotquot volvit uterque polus;
Collaque bis vivi Pelopis quae brachia vincant,
  Quaeque fluit puro nectare tincta via,
Et decus eximium frontis, tremulosque capillos,
  Aurea quae fallax retia tendit Amor.   60
Pellacesque genas, ad quas hyacinthina sordet
  Purpura, et ipse tui floris, Adoni, rubor.
Cedite laudatae toties Heroides olim,
  Et quaecunque vagum cepit amica Iovem.
Cedite Achaemeniae turrita fronte puellae,   65
  Et quot Susa colunt, Memnoniamque Ninon.
Vos etiam Danaae fasces submittite Nymphae,
  Et vos Iliacae, Romuleaeque nurus.
Nec Pompeianas Tarpeia Musa columnas
  Iactet, et Ausoniis plena theatra stolis.   70
Gloria virginibus debetur prima Britannis,
  Extera sat tibi sit foemina posse sequi.

Tuque urbs Dardaniis Londinum structa colonis
  Turrigerum late conspicienda caput,
Tu nimium felix intra tua moenia claudis   75
  Quicquid formosi pendulus orbis habet.

54 senium (old age) M's youthful concern with senility. Cf. regeneration of other parent-figures in *El V*, and *El VI* 55, 'adult Jove'. **55 lumina** basic meaning lights, usually artificial, also refers to eyes, a double meaning M often used. **57 Pelopis** (Pelops') Tantalus killed Pelops, his son, and served him in a feast for the gods. Only Demeter ate anything, and Pelops was restored to life lacking a shoulder, which the gods replaced with ivory. M's point in using this myth isn't clear. **59 decus** vague word with anti-erotic connotations of decency, respectability. **tremulos** either shimmering (of light) or waving in the wind – or both. **62 Adoni** (Adonis) beloved of Venus, the anemone sprang from his blood (Ovid *Met* x 731). **65 turrita** (turreted) referring to the high hats ancient Persian women were said to have worn as signs of rank, later taken as symbols of their degenerate love of display. Perhaps also the high head-dresses worn by fashionable women in 17c. Cybele, a fertility goddess, wore a towered front (see *El V* 62). This may be mere literary padding though. **66 Memnoniam** (Memnon's) Memnon founded Susa but not Nineveh.

How often I've stood, struck dumb at marvels of beauty
    which might have restored a senile Jupiter's lust.
How often I've gazed on eyes that were brighter than
        jewels,   55
    than all the heavenly torches in either pole,
and necks that were whiter than Pelops' ivory arms,
    and the milky way, which flows with pure nectar:
superbly gracious foreheads, and shimmering hair,
    those golden nets where Love catches his prey: 60
seductive cheeks, which make a hyacinth's purple
    and even the blush of your flower, Adonis, seem dull.
You beauties famous in legend, admit they surpass you,
    and every girl beloved of amorous Jove;
you too, you maidens of Persia with towering head-
        dresses,   65
    and Assyrian girls from Susa and Nineveh.
Acknowledge yourselves inferior, you nymphs of Greece,
    and daughters of Troy, and beauties of ancient Rome.
And Ovid, don't boast of Pompeian pillars of beauty,
    or theatres filled with the ample blouses of Rome.   70
The highest honour belongs to the British girls:
    you foreign women, aim to be second at best.

And London! you city that Trojan colonists founded,
    your turreted head is visible far and wide;
lucky beyond all measure your walls shut in   75
    whatever beauty this pendulous orb contains.

M may not quite know where he is, but it probably doesn't matter.
69 Tarpeia (to do with the Tarpeian rock, in Rome, near where Ovid
lived). Consistent transformation of people into things here. Ovid praised
not the columns but the girls who stood in front of them – as still as
columns? The theatre is full of blouses, **stolis**, an upper garment worn
by Roman matrons, but Ovid was interested in what filled the blouse.
plena (full) probably a joke along these lines.   **73 Dardaniis** (Trojan)
referring to the legend that Britain was originally colonized by Brutus,
a Trojan. Hence it was built by exiles (like himself) and, defeating the
point of his comparison, British women were all descended from foreign-
ers.   **74 turrigerum** (tower-bearing) London at this time had many
vertical structures, especially church towers. Connects with *turrita* at
65, and perhaps suggests some criticism of London's ostentation, in-
consistently with his apparent point here but in keeping with 87.

Non tibi tot caelo scintillant astra sereno
    Endymioneae turba ministra deae,
Quot tibi conspicuae formaque auroque puellae
    Per medias radiant turba videnda vias,   80
Creditur huc geminis venisse invecta columbis
    Alma pharetrigero milite cincta Venus,
Huic Cnidon, et riguas Simoentis flumine valles,
    Huic Paphon, et roseam posthabitura Cypron.

Ast ego, dum pueri sinit indulgentia caeci,   85
    Moenia quam subito linquere fausta paro;
Et vitare procul malefidae infamia Circes
    Atria, divini Molyos usus ope.
Stat quoque iuncosas Cami remeare paludes,
    Atque iterum raucae murmur adire Scholae.   90
Interea fidi parvum cape munus amici,
    Paucaque in alternos verba coacta modos.

82 **Alma pharetrigero** (bountiful, kind, with quiver-bearing troops) the
dual nature of Venus, not otherwise prominent in this poem, but perhaps
preparing for the switch at 87.   83 **Cnidon** (Cnidos) etc places associated
with Venus.   87 **Circes** daughter of the Sun (*Comus* 51) symbolic of
sensual indulgence, turned Ulysses' sailors into swine. Ulysses escaped,
protected by the moly plant (Homer *Odyssey* 10). A total switch of

The stars that sparkle for you from a tranquil sky,
  the throng attending the moon, Endymion's love,
fall short of the maidens brilliant with beauty and gold
  who draw all eyes as they blast through your streets
      like light.   80
Bountiful Venus, borne by twin doves it is said
  came hither, ringed by her troops with their quivers full,
despising Cnidos, and valleys Simóis flows through,
  preferring London to Paphos and rosy Cyprus.

But while blind Cupid still kindly leaves me unscathed,   85
  I prepare to swiftly leave these happy walls,
and fly the infamous halls of treacherous Circe,
  using the power of the magic Moly plant.
My grand return to the reeds and mud of the Cam
  is fixed: back to the creaking classroom I go.   90
So accept this meagre gift a faithful friend,
  a few words cramped into metrical form.

orientation. M is now the heroic Ulysses, fighting off temptations as he
returns home (to Cambridge?). **89 remeare** (return) particularly used
of triumphal returns, so mock-heroic image for his grand return to
Cambridge. **92 alternos** (alternate) the alternating hexameters and
pentameters of the elegiac form.

Elegia V   In adventum Veris

I N SE PERPETUO Tempus revolubile gyro
   Iam revocat zephyros vere tepente novos.
  Induiturque brevem Tellus reparata iuventam,
   Iamque soluta gelu dulce virescit humus.
Fallor? an et nobis redeunt in carmina vires,   5
   Ingeniumque mihi munere veris adest?
Munere veris adest, iterumque vigescit ab illo
   (Quis putet) atque aliquod iam sibi poscit opus.

Castalis ante oculos, bifidumque cacumen oberrat,
   Et mihi Pyrenen somnia nocte ferunt.   10
Concitaque arcano fervent mihi pectora motu,
   Et furor, et sonitus me sacer intus agit.
Delius ipse venit, video Peneide lauro
   Implicitos crines, Delius ipse venit.
Iam mihi mens liquidi raptatur in ardua coeli,   15
   Perque vagas nubes corpore liber eo.
Perque umbras, perque antra feror penetralia vatum,
   Et mihi fana patent interiora Deum.
Intuiturque animus toto quid agatur Olympo,
   Nec fugiunt oculos Tartara caeca meos.   20
Quid tam grande sonat distento spiritus ore?
   Quid parit haec rabies, quid sacer iste furor?

1 **perpetuo** (perpetual) Time's timelessness, juxtaposed with *Tempus*, time.
Note all the compounds with *re*, again, emphasizing the cyclic nature of
time.   3 **Induitur** (put on clothing) a very external image to use of this
process.   5 **Fallor** (Am I mistaken?) a rhetorical mannerism from Ovid.
9 **cacumen** (peak) Parnassus, like the Pirenian spring, sacred to poetry.
Pegasus, winged horse also symbolic of poetry, drank at the spring of
Pirene. This mobile geography is the first sign to M of his poetic in-
spiration; later he will be airborne himself.   13 **Delius** (Apollo, from
his birthplace, Delos). Daphne, who was transformed into laurel to
escape the lust of the god, is referred to by reference to her father
Peneus' name.   14 **implicitos** (woven into) the hair of Apollo,
symbolic of the sun's rays, tangled with Daphne's laurel (sexual
attractions) – but Apollo here is unworried, in contrast to *El VII*, and
unlike Samson.   17 **umbras** (shadows; also shades of the dead, and

Elegy V On the coming of spring

TIME THAT REVOLVES for ever around itself
    recalls new zephyrs, as glowing spring returns.
    The earth is restored, and clothed in transient
        youth,
the soil is freed from the frost, and sweetly greens.
And unless I'm deceived, the vigour returns to my verse,   5
    my natural talents emerge as the gift of spring.
Yes! as the gift of spring they flourish again –
    amazing! – and now demand to be given a task.

Castalia's fount and Parnassus' twin peaks confront
    me,
    and at night my slumbers bring me the Pirenian   10
    spring.
My pounding heart is on fire with secret feelings,
    impelled by a holy madness and tumult within.
Apollo himself is coming! – I see his curls
    encircled with Daphne's laurel – Apollo comes!
My mind is swept to the heights of the liquid sky,   15
    and freed from my body I pass through wandering
    clouds.
I'm borne through shades, through the innermost shrines
    of the bards,
    and the gods' most secret places lie open before me.
My soul at a glance sees all that is done on Olympus,
    and nothing in murky hell can escape my gaze.   20
And what does my spirit thunder with mouth straining?
    What fruit does this madness, this sacred frenzy bring
    forth?

images or replicas of ideas). The ambiguities of the word help make
the transition from M's bodiless ecstasy to his very physical vision.
20 caeca basic meaning blind, which the inhabitants of Hades are, but
could mean defeating sight, ie unseeable. **22 parit** (gives birth to)
connects his poetic creativity with the fertility of animal mothers,
especially in spring.

Ver mihi, quod dedit ingenium, cantabitur illo;
Profuerint isto reddita dona modo.

Iam Philomela tuos foliis adoperta novellis   25
Instituis modulos, dum silet omne nemus.
Urbe ego, tu sylva simul incipiamus utrique,
Et simul adventum veris uterque canat.
Veris io rediere vices, celebremus honores
Veris, et hoc subeat Musa perennis opus.   30
Iam sol Aethiopas fugiens Tithoniaque arva,
Flectit ad Arctoas aurea lora plagas.
Est breve noctis iter, brevis est mora noctis opacae
Horrida cum tenebris exulat illa suis.
Iamque Lycanoius plaustrum caeleste Bootes   35
Non longa sequitur fessus ut ante via,
Nunc etiam solitas circum Iovis atria toto
Excubias agitant sydera rara polo.
Nam dolus, et caedes, et vis cum nocte recessit,
Neve Giganteum Dii timuere scelus.   40

Forte aliquis scopuli recubans in vertice pastor,
Roscida cum primo sole rubescit humus,
'Hac,' ait, 'hac certe caruisti nocte puella
Phoebe tua, celeres quae retineret equos.'
Laeta suas repetit sylvas, pharetramque resumit   45
Cynthia, Luciferas ut videt alta rotas,
Et tenues ponens radios gaudere videtur
Officium fieri tam breve fratris ope.

25 **Philomela** nightingale, from Greek meaning lover of song. **foliis**
(leaves) could refer also to sheets of paper. 26 **Instituis** (get ready)
Philomena like M is only planning her song at this point. The grove
is still silent. 30 **perennis** (perennial, everlasting) an eternal Muse
for a recurring rite. In 1645 M had *quotannis* (of every year) which
makes better sense but didn't scan. M was criticized by Salmasius
for an error in his Latin, and changed it for the 1673 edition. 31
Aethiopas (Aethiopians) referring to the equator, which M thought
passed through Aethiopia (cf *PL* iv 282). Tithonus, husband of the
dawn, represents the east. M is describing the altered course of the

Spring, which inspired my song, will provide my
     theme.
His gifts repaid in such measures will be his reward.

Philomela, shrouded in pristine leaves you settle   25
     to work on your songs, while all the grove is still.
Let's start at a single instant, I in the city,
     you in the woods, to sing the coming of spring.
Hurrah for the spring's return! Let's honour the rites
     of spring: may the never-changing Muse assist.   30
And the sun now flees from the tropics, and Tithonus'
     fields,
     and turns his golden reins towards the north.
Short is the journey of shadowy night, and short
     her stay: horrid night and her gloom are banished.
The Bear in the north is no longer weary from trailing   35
     the Heavenly Waggon's protracted nightly course;
and only a handful of stars is left in the sky
     to keep their accustomed watch round Jupiter's
     courts,
for treachery, murder and violence have gone with the
     night,
     and the gods no longer fear a giants' revolt.   40

Some shepherd perhaps, reclining on top of a crag
     while the dewy ground grows red with the early sun,
may say: 'Apollo, last night I'm sure you were missing
     a girl to hold your eager horses in check.'
And Cynthia, donning her quiver, returns with joy   45
     to her woods, gazing down on the sun's bright
     wheels,
and sheathing her tenuous rays she seems to rejoice
     that thanks to her brother her task becomes so short.

sun at the vernal equinox. **41 recubans** (reclining) picks up *Excubias*
from 38. What does this shepherd represent? Does he foreshadow the
hunter of 52 on Hymettus? **43 puella** (girl) what girl is this? **46**
Luciferas (light-bearing) of the wheels of the sun's chariot, but Lucifer
was also a name for Satan.

'Desere', Phoebus ait, 'thalamos Aurora seniles,
    Quid iuvat effoeto procubuisse toro?   50
Te manet Aeolides viridi venator in herba,
    Surge, tuos ignes altus Hymettus habet.'
Flava verecundo dea crimen in ore fatetur,
    Et matutinos ocyus urget equos.'

Exuit invisam Tellus rediviva senectam,   55
    Et cupit amplexus Phoebe subire tuos;
Et cupit, et digna est, quid enim formosius illa,
    Pandit ut omniferos luxuriosa sinus,
Atque Arabum spirat messes, et ab ore venusto
    Mitia cum Paphiis fundit amoma rosis.   60
Ecce coronatur sacro frons ardua luco,
    Cingit ut Idaeam pinea turris Opim;
Et vario madidos intexit flore capillos,
    Floribus et visa est posse placere suis.
Floribus effusos ut erat redimita capillos   65
    Tenario placuit diva Sicana Deo.

Aspice Phoebe tibi faciles hortantur amores,
    Mellitasque movent flamina verna preces.
Cinnamea Zephyrus leve plaudit odorifer ala,
    Blanditiasque tibi ferre videntur aves.   70
Nec sine dote tuos temeraria quaerit amores
    Terra, nec optatos poscit egena toros,

49 thalamos (bridal chamber) refers to Tithonus, human husband of
Dawn, granted immortality by the gods, but not eternal youth, so he
withered away. M's bold image links the bridal chamber of his youth
with his present old age, and by calling him a room suggests that he
is no longer a person. So also *toro* (couch) in the next line. **51
Aeolides** (descendant of Aeolus). Cephalus was either son or grandson of Aeolus. The point of calling him this is to emphasize his
status as a son. **venator** (hunter) recalls the unhappy part of the
story; Cephalus accidentally killed his loving wife Procris who was
concealed in the long grass. **52 ignes** (fires) double meaning, referring to Dawn's object of love, Cephalus, and also to the fires of
Dawn striking Hymettus. **57 quid** (what) neuter, so the feminine earth
is being compared to inanimate things. A break in M's metaphor?
58 sinus (folds, especially of the cloth over the breast or lap). So the

'Dawn', says Apollo, 'abandon that senile bedroom –
    What joy can come from pressing a barren couch?   50
Aeolus' grandson, a hunter, waits in the green grass;
    get up, for steep Hymettus holds your flame.'
The golden goddess' blushes confess her guilt,
    and she urges the horses of morn to quicken their pace.

And Earth is reborn, discarding contemptible age,   55
    and yearns to submit, Apollo, to your embrace.
She yearns, and is worthy; for what is more lovely than her,
    wantonly flaunting the breasts of the mother of all?
and breathing Arabian spices, and pouring sweet balm
    mingled with roses of Paphos from lovely lips?   60
Her towering crest is crowned with a sacred grove,
    like the tower of pines that circles Idaean Ops.
And she plaits her sodden curls with flowers of all kinds,
    flowers of her own with manifest power to please.
Divine Persephone wreathed her dishevelled hair   65
    with flowers like these, to please the god of the dead.

Look round you, Apollo, for yielding loves entice you,
    breezes of spring are wafting honeyed prayers;
the scented zephyr rustles his cinnamon wings;
    for you it seems the birds bring flattering songs.   70
And she asks for your love with decorum, and brings a
        dowry:
    earth is no pauper to beg for your happy bed.

word M uses is vague and unspecific, but *luxuriosa* (sensuous) immediately
before it insists on her sensual motives. **62 Opim** (Ops) fertility goddess, wife of Saturn. M links her with Cybele, mother of the gods,
through the tower-image associated with Cybele. M seems unclear
about the relationship between these three figures, Earth and the two
goddesses. **63 madidos** (drenched) as with wine – image of over-
whelming fecundity, but not especially winning if applied to a human
girl. **64 visa** (is seen or seems) M's point here isn't clear. **66 Tenario**
of Taenarus, in Greece; traditionally the entrance to Hades. M refers
here to the story of Proserpina and Pluto. See also *PL* IV 258. **70
videntur** (seem or are seen to) as in 64 not clear whether M is stressing
the visibility of the process, or contrasting appearances with the reality,
that the birds are singing for their own loves, not to Apollo.

Alma salutiferum medicos tibi gramen in usus
Praebet, et hinc titulos adiuvat ipsa tuos.
Quod si te pretium, si te fulgentia tangunt   75
Munera (muneribus saepe coemptus Amor),
Illa tibi ostentat quascunque sub aequore vasto,
Et superiniectis montibus abdit opes.

Ah quoties cum tu clivoso fessus Olympo
In vespertinas praecipitaris aquas,   80
'Cur te', inquit, 'cursu languentem Phoebe diurno
Hesperiis recipit caerula mater aquis?
Quid tibi cum Tethy? Quid cum Tartesside lympha,
Dia quid immundo perluis ora salo?
Frigora Phoebe mea melius captabis in umbra,   85
Huc ades, ardentes imbue rore comas.
Mollior egelida veniet tibi somnus in herba,
Huc ades, et gremio lumina pone meo.
Quaque iaces circum mulcebit lene susurrans
Aura per humentes corpora fusa rosas.   90
Nec me (crede mihi) terrent Semeleia fata,
Nec Phaetonteo fumidus axis equo;
Cum tu Phoebe tuo sapientius uteris igni,
Huc ades et gremio lumina pone meo.'

Sic Tellus lasciva suos suspirat amores;   95
Matris in exemplum caetera turba ruunt.
Nunc etenim toto currit vagus orbe Cupido,
Languentesque fovet solis ab igne faces.
Insonuere novis lethalia cornua nervis,
Triste micant ferro tela corusca novo.   100

75 pretium (price) strange insistence on the mercenary nature of love.
79 clivoso (steep) image is unclear, whether M means the mountain, *on*
which Apollo is briefly resting before the final descent, or the sky, *from*
which he is descending. 'Steep' suggests the former, which would be more
suitable thematically: Apollo resting on a mountain like Cephalus on
Hymettus. **83 Tethy** mother of rivers. **Tartesside** (daughter of
Tartessus) the Atlantic. **85 Frigora** (coldness, chill) sounds ominous.
88 lumina (eyes/brightness) see Intro on this pun. **91 Semeleia**
(Semele's). Beloved of Jove, she asked to see him in his divine form

She bountifully offers you health-giving herbs to use
    as medicine, thus enhancing your fame as healer.
And if money is what you want, if you're moved by   75
    glittering
    gifts (for love is often bought with gifts),
she'll show you uncountable riches hidden beneath
    the boundless ocean and mountains piled on high.

And often she's asked, as you wearily plummet down
    from steep Olympus towards the evening waves: 80
'Apollo, why should the azure mother receive you,
    faint from your daily round, in the western waves?
Do sea-nymphs concern you? Or Tethys the mother of
    waters?
Why sluice the face of a god in filthy brine?
Better to seek for coolness in shades on earth: 85
    come hither and drench your blazing curls in dew.
You'll sleep more softly, lying on chilly grass:
    come hither and lay your brightness in my lap.
Wherever you lie, a gently whispering breeze
    will caress our bodies, sprawling on moistened roses.   90
The fate of Semele frightens me not at all,
    nor Phaeton's horses, who whirled the axle to smoke.
So learn to direct your fires more wisely, Apollo:
    come hither and lay your brightness in my lap.'

Thus sighs lascivious earth, proclaiming her love,   95
    and the rest of the rabble follow their mother's lead.
Promiscuous Cupid ranges all over the globe,
    and kindles his drooping torch with fire from the sun;
and his horn-bow sings, deadly with new-stung gut,
    and his bright darts glitter with cruel new barbs.   100

and was destroyed by the flame. Her child by Jove was Dionysios.
92 Phaetonteo (Phaeton's) son of Apollo, asked to be allowed to drive
the chariot of the sun, but died when he lost control of the horses (a son
less powerful than his father, unusually for this poem). **93 sapientius**
(more wisely) the prudence of passion. **97 orbe** (globe) the earth, de-
personified now. **99 cornua** (horns) of Cupid's bow, but probably a
sexual sense as well.

Iamque vel invictam tentat superasse Dianam,
    Quaeque sedet sacro Vesta pudica foco.
Ipsa senescentem reparat Venus annua formam,
    Atque iterum tepido creditur orta mari.

Marmoreas iuvenes clamant 'Hymenaee' per urbes,   105
    Litus 'io Hymen', et cava saxa sonant.
Cultior ille venit tunicaque decentior apta,
    Puniceum redolet vestis odora crocum.
Egrediturque frequens ad amoeni gaudia veris
    Virgineos auro cincta puella sinus.   110
Votum est cuique suum, votum est tamen omnibus unum,
    Ut sibi quem cupiat, det Cytherea virum.
Nunc quoque septena modulatur arundine pastor,
    Et sua quae iungat carmina Phyllis habet.
Navita nocturno placat sua sidera cantu,   115
    Delphinasque leves ad vada summa vocat.

Iupiter ipse alto cum coniuge ludit Olympo,
    Convocat et famulos ad sua festa Deos.
Nunc etiam Satyri cum sera crepuscula surgunt,
    Pervolitant celeri florea rura choro,   120
Sylvanusque sua cyparissi fronde revinctus,
    Semicaperque Deus, semideusque caper.
Quaeque sub arboribus Dryades latuere vetustis
    Per iuga, per solos expatiantur agros.
Per sata luxuriat fruticetaque Maenalius Pan,   125
    Vix Cybele mater, vix sibi tuta Ceres,
Atque aliquam cupidus praedatur Oreada Faunus,
    Consulit in trepidos dum sibi nympha pedes,
Iamque latet, latitansque cupit male tecta videri,
    Et fugit, et fugiens pervelit ipsa capi.   130

102 **Vesta** daughter of Ops, sister of Ceres, her shrine in Rome was tended
by the Vestal virgins.   106 **'io Hymen'** ritual shout to greet Hymen, god
of marriage.   115 **nocturno** (nocturnal) day has been passing as the poem
progresses. **sidera** (stars) perhaps the eyes of his shepherdess. 121 **cyparissi**
(cypress) a boy loved by Sylvanus, turned into the cypress, which Sylvanus
carried in memory of him.   122 **Semicaperque** (half-goatish) recalling a

He tries to conquer even Diana, the invincible,
  and Vesta, who chastely sits by the sacred shrine.
And year-worn Venus restores her aging beauty,
  and seems to have risen again from the warm sea.

Through cities of marble the young men shout out   105
  'Hymen!'
  and shores and hollow rocks resound with 'Hymen!'
He comes, discreetly adorned, in ritual tunic,
  his scented garments smelling of purple crocus.
And girls go forth for the pleasant delights of spring
  in throngs, with golden clasps round virgin breasts.   110
Each has a prayer of her own, but their prayers are the same –
  that Venus will give each one the man she desires.
And a shepherd is playing a tune on his seven-reed pipe,
  and Phyllis has songs of her own which marry his.
The sailor placates his stars with a nightly song,   115
  and calls on agile dolphins to leap from the waves.

Great Jupiter sports with his wife on high Olympus,
  inviting even menial gods to his feasts.
And the Satyrs now, with twilight flourishing late,
  dance through flowery fields in a swift-footed band.   120
Sylvanus as well, wreathed again with his cypress,
  the god who is half a goat, the goatish demi-god.
And Dryads who hid before under ancient trees
  now roam abroad on mountains and lonely fields.
And Arcadian Pan runs riot through crops and thickets,   125
  and mother Cybéle is hardly safe, nor Ceres.
And randy Faunus catches a mountain nymph,
  but the nymph resolves to trust to her trembling feet,
and hides, but is poorly concealed, for she wants to be seen,
  and flees, but fleeing she really wants to be caught.   130

line by Ovid famous in antiquity for showing lack of taste, so the image
here isn't meant to be taken fully seriously. **125 sata** (crops) **fruticeta**
(bush) cultivated and uncultivated nature. Cybele in the next line may be
the source of natural energies in general (hence related to *fruticeta*) and
Ceres of crops (*sata*). **128** the sense of this is clear, but M's Latin has
been objected to.

Dii quoque non dubitant caelo praeponere sylvas,
  Et sua quisque sibi numina lucus habet.

Et sua quisque diu sibi numina lucus habeto,
  Nec vos arborea dii precor ite domo.
Te referant miseris te Iupiter aurea terris   135
  Saecla, quid ad nimbos aspera tela redis?
Tu saltem lente rapidos age Phoebe iugales
  Qua potes, et sensim tempora veris eant.
Brumaque productas tarde ferat hispida noctes,
  Ingruat et nostro serior umbra polo.   140

135 **aurea** (golden) instead of a god bringing a golden age, M has this
golden age bringing back a god, a paradox whose falsity has already been
brought out by **miseris** (wretched). 139 **hispida** (shaggy) winter like

And the gods are sure they prefer these woods to heaven,
and every sacred grove has its own godhead.

And long may every grove have its own godhead!
I pray you'll never leave your arboreal homes.
Let the golden times recall you to suffering earth,   135
Jupiter, not back to your clouds and fierce thunders.
Or Apollo, at least you could drive your fleet-footed team
as slow as you can, to delay the passing of spring.
May shaggy winter be late, with its lengthening nights,
and shadows slower to fall on our northern skies.   140

the coats of animals grown for protection. So winter is rough, uncouth,
but hair for M is always a mark of vitality.

Elegia VII

NONDUM BLANDA TUAS LEGES Amathusia noram,
    Et Paphio vacuum pectus ab igne fuit.
    Saepe cupidineas, puerilia tela, sagittas,
Atque tuum sprevi maxime, numen, Amor.
'Tu puer imbelles' dixi 'transfige columbas,   5
    Conveniunt tenero mollia bella duci.
Aut de passeribus tumidos age, parve, triumphos,
    Haec sunt militiae digna trophaea tuae:
In genus humanum quid inania dirigis arma?
    Non valet in fortes ista pharetra viros.'   10
Non tulit hoc Cyprius (neque enim deus ullus ad iras
    Promptior) et duplici iam ferus igne calet.

Ver erat, et summae radians per culmina villae
    Attulerat primam lux tibi Maie diem:
At mihi adhuc refugam quaerebant lumina noctem   15
    Nec matutinum sustinuere iubar.
Astat Amor lecto, pictis Amor impiger alis,
    Prodidit astantem mota pharetra deum:
Prodidit et facies, et dulce minantis ocelli,
    Et quicquid puero, dignum et Amore fuit.   20
Talis in aeterno iuvenis Sigeius Olympo
    Miscet amatori pocula plena Iovi;
Aut qui formosas pellexit ad oscula nymphas
    Thiodamantaeus Naiade raptus Hylas;

1 see Commentary p. 8. 4 **Amor** (love) change of sex for love, from Venus to Cupid. 5 **columbas** (doves) birds associated with Venus. 7 **parve** (small one) carefully placed to puncture the hyperbole. 9 **humanum** (human) sees sex as only for animals. 10 **viros** (males, husbands) ironic, since the word refers to sexually mature and active males, which M isn't, in spite of his jibe at Cupid. 11 **Cyprius** (Cyprian) usually referring only to Venus. 12 **duplici** (double) since fire has an ambiguous reference, either to rage or sexual passion, this could mean either that he was twice as angry, or that he was building up twice the sexual passion for M (on the principle that the longer one resists love, the more intensely one feels it finally. See *Damon* 86). 17 **impiger** (indefatigable) so is M. 20 **dignum** (worthy) non-erotic word about

Elegy VII

VENUS, SWEET TEMPTRESS, before I knew of your laws,
    my heart still free, untouched by Paphian fires,
I often sneered at Cupid's childish weapons,
    his arrows, and loudly scorned your godhead, Love.
'You child,' I said, 'Go shoot at some harmless doves,   5
    an easy battle, fit for a fledgling chief!
Or order a full-blown triumph for conquering sparrows:
    trivial victories won in a paltry campaign.
Why aim your frivolous darts at the human race?
    Your arrows will not prevail with powerful males.'   10
Venus' son was offended (no god is prompter
    to wrath): he fiercely burns with a double flame.

It was spring, and the sunlight gleamed on the roofs of the tallest
    houses, to bring you the first of your month, O May.
But my eyes like lamps still looked to vanishing night,   15
    unable to stand the brilliance of the dawn.
And industrious Love with his painted wings stood close
    to my bed, his presence betrayed by rustling arrows,
betrayed by his face, by his sweetly menacing eyes,
    and all the rest that graces a boy, or Love: 20
like Ganymede, young in Olympus' eternal halls,
    mixing the brimming bowls for his lover, Jove,
or the youth whose charms drew beautiful nymphs to kiss him,
    Hylas, Thiódamus' son, whom a Naiad raped.

erotic charms. **21 aeterno** (eternal) contrasted with the youth of
Ganymede (*iuvenis*). **22 pocula** (cups) Ganymede became cup-bearer
to the gods. The mention here suggests the union in heaven of love and
wine (Dionysios). **23 formosas** (beautiful) of the nymph, linking her
to Hylas who is also beautiful. M's language makes Hylas sound more
active than he was. One of the Argonauts, boyfriend of Hercules, he
was the passive victim of some water-nymphs, who drew him into their
waters and drowned him in their desire to make love to him (see
Propertius I 20).

Addideratque iras, sed et has decuisse putares, 25
Addideratque truces, nec sine felle minas.

'Et miser exemplo sapuisses tutius', inquit,
'Nunc mea quid possit dextera testis eris.
Inter et expertos vires numerabere nostras,
Et faciam vero per tua damna fidem. 30
Ipse ego si nescis strato Pythone superbum
Edomui Phoebum, cessit et ille mihi;
Et quoties meminit Peneidos, ipse fatetur
Certius et gravius tela nocere mea.
Me nequit adductum curvare peritius arcum, 35
Qui post terga solet vincere Parthus eques.
Cydoniusque mihi cedit venator, et ille
Inscius uxori qui necis author erat.
Est etiam nobis ingens quoque victus Orion,
Herculeaeque manus, Herculeusque comes. 40
Iupiter ipse licet sua fulmina torqueat in me,
Haerebunt lateri spicula nostra Iovis.
Caetera quae dubitas melius mea tela docebunt,
Et tua non leviter corda petenda mihi.
Nec te stulte tuae poterunt defendere musae, 45
Nec tibi Phoebaeus porriget anguis opem.'

Dixit, et aurato quatiens mucrone sagittam,
Evolat in tepidos Cypridos ille sinus.
At mihi risuro tonuit ferus ore minaci,
Et mihi de puero non metus ullus erat, 50

31 **Pythone** (Pytho) M gave this incident a political gloss in 1642, predicting the end of prelacy; 'till like that fen-born serpent she be shot to death with darts of the sun, the pure and powerful beams of God's word' (*Reason of church government*). 33 **Peneidos** (daughter of Peneus) Daphne (see *El V* 12 and Commentary p. 17). 36 **post terga** (from behind) Parthian cavalry was greatly respected by the Romans. M brings out the significance of this particular tactic in *PR* III 325: they shoot on the run, and 'overcome by flight'. Here probably a reference to Daphne's backward glances as she fled, which inflamed Apollo even more. 37 **Cydonius** (Cretan) perhaps Hippolytus, who was Cretan and a hunter (see Euripides *H* 18). **ille** (the one)

74

Like them, but angry as well (but charmingly so)   25
and fierce were his threats, tinged with a genuine rage.

'Poor fool', he said, 'It's safer to learn by example,
but now you'll see directly the power of my arm.
You'll count as one of the ones who has felt my power:
your sufferings will add believers to my cult.   30
It was me – if you didn't know it – who tamed Apollo,
arrogant slayer of Pytho: he yielded to me,
and whenever he thinks of Daphne, himself agrees
my arrows are surer and give a deeper wound.
I can draw a tight-strung bow with skill that excels   35
the Parthian horseman's, who shoots from behind to win.
The Cretan hunter defers to my skill, and Cephalus
too, unwitting cause of the death of his wife.
And mighty Orion as well was conquered by me,
and Hercules' masculine hands, and Hercules' friend.   40
If Jupiter chose to turn his thunder against me,
my barbs would stick in the sides of Jove himself.
If doubts remain, my darts will teach better than words;
and your heart, the target I have to hit, and hard.
Those Muses of yours will be powerless to help you,   45
fool,
and Apollo's healing serpent will lend you no aid.'

He ended, and waving an arrow with gilded tip
he wafted away to the warm breast of Venus.
But the sight of him fiercely thundering threats seemed
laughable
then, and I felt not the slightest fear of the boy.   50

Cephalus. See Commentary pp. 17–18. **40 manus** (hands) Hercules in
love with Omphale was made by her to dress in women's clothes and
spin thread. In Ovid's *Heroides*, Deianira his wife taunts him about it,
and especially the clumsiness of his hands in this inappropriate task.
comes (friend or boyfriend) could refer to one of Hercules' many
friends who was enslaved by love, like Jason or Theseus, or a boyfriend, like Hylas. **44 petenda** (must be sought out) Cupid suggests
some reluctance here, as though regretful that he must hurt M's heart
(which is his domain). **46 anguis** (snake) symbol of healing.

Et modo qua nostri spatiantur in urbe quirites
Et modo villarum proxima rura placent.
Turba frequens, facieque simillima turba dearum
Splendida per medias itque reditque vias.
Auctaque luce dies gemino fulgore coruscat,   55
Fallor? an et radios hinc quoque Phoebus
      habet.

Haec ego non fugi spectacula grata severus,
Impetus et quo me fert iuvenilis, agor.
Lumina luminibus male providus obvia misi
Neve oculos potui continuisse meos.   60
Unam forte aliis supereminuisse notabam,
Principium nostri lux erat illa mali.
Sic Venus optaret mortalibus ipsa videri,
Sic regina deum conspicienda fuit.
Hanc memor obiecit nobis malus ille Cupido,   65
Solus et hos nobis texuit ante dolos.
Nec procul ipse vafer latuit, multaeque sagittae,
Et facis a tergo grande pependit onus.
Nec mora, nunc ciliis haesit, nunc virginis ori,
Insilit hinc labiis, insidet inde genis: 70
Et quascunque agilis partes iaculator oberrat,
Hei mihi, mille locis pectus inerme ferit.
Protinus insoliti subierunt corda furores,
Uror amans intus, flammaque totus eram.

Interea misero quae iam mihi sola placebat,   75
Ablata est oculis non reditura meis.

51 quirites (citizens of Rome) ordinary Londoners, not lower class.
52 villarum (estates or towns) plural but perhaps singular in significance,
referring to a particular estate M visited. See *El I* 50.   **55 gemino** (twin)
these two lines suggest that Cupid and Apollo are no longer antagonists,
but complementary powers. Apollo's energies are erotic – perhaps.   **57
grata** (pleasing). **severus** (stern) juxtaposing pleasure with his puritanism.
58 fert (bears). **agor** (I am driven) M's sense of passivity.   **59 Lumina**
(lights/eyes) another use of this important pun.   **61 forte** (by chance) in fact
by Cupid's design.   **62 lux** (light) either her beauty, or the sight of
her, or the light of that day. The line has an ominous generality, later

And sometimes I stroll in the city where townsfolk parade,
and sometimes delight in the grounds of noble estates.
And a mighty throng with faces as lovely as goddesses'
comes and goes in splendour along the paths.
And the day dazzles, its brilliance doubled by theirs –   55
perhaps the rays of Apollo come from here?

I'm not so strict that I'd shun so pleasant a show,
I'm driven wherever youthful passions take me.
And reckless my gaze met theirs, my brightness their
brightness,
I was quite unable to hold my eyes in check.   60
Then one I happened to notice, excelling the rest,
her radiance then the root of ills to come.
Venus would want to look thus, appearing to mortals;
she drew the eyes of all, like the queen of the gods.
Patient malevolent Cupid had thrust her before me,   65
alone he wove the snares, prepared in advance.
And the prankster was hiding nearby, with hundreds of
arrows,
his torch drooping behind him, a heavy weight.
No lingering; he clings to the eyelids and mouth of the
girl,
then darts between her lips, then camps on her cheek.   70
And wherever the agile archer roams on her person,
he wounds my helpless breast with thousands of darts.
Immediately, strange emotions invaded my heart,
I burn within with love, turned wholly to flame.

Meanwhile the one who alone gave joy to my pain   75
had gone, never again to return to my eyes.

to be echoed in *PL* 1 1 'the fruit Of that forbidden tree, whose mortal
taste Brought death into the world, and all our woe'. **nostri** could be
either singular in sense (my), or plural (our). **66 ante** (before) either
of place, in front of my feet, or time, beforehand. **72 locis** (places)
either from a thousand places, as he dances round, or in a thousand
places in my heart. The two possibilities reinforce each other. **74 eram**
(I was) tense shift to past, out of the immediacy of the experience.

Ast ego progredior tacite querebundus, et excors,
    Et dubius volui saepe referre pedem.
Findor, et haec remanet, sequitur pars altera votum,
    Raptaque tam subito gaudia flere iuvat.   80
Sic dolet amissum proles Iunonia coelum,
    Inter Lemniacos praecipitata focos.
Talis et abreptum solem respexit, ad Orcum
    Vectus ab attonitis Amphiaraus equis.

Quid faciam infelix, et luctu victus, amores   85
    Nec licet inceptos ponere, neve sequi.
O utinam spectare semel mihi detur amatos
    Vultus, et coram tristia verba loqui!
Forsitan et duro non est adamante creata,
    Forte nec ad nostras surdeat illa preces.   90
Crede mihi nullus sic infeliciter arsit,
    Ponar in exemplo primus et unus ego.
Parce precor teneri cum sis deus ales amoris,
    Pugnent officio nec tua facta tuo.
Iam tuus O certe est mihi formidabilis arcus,   95
    Nate dea, iaculis nec minus igne potens:
Et tua fumabunt nostris altaria donis,
    Solus et in superis tu mihi summus eris.
Deme meos tandem, verum nec deme furores,
    Nescio cur, miser est suaviter ominis amans: 100
Tu modo da facilis, posthaec mea siqua futura est,
    Cuspis amaturos figat ut una duos.

77 **progredior** (advance) where to? Perhaps metaphorically, of his quest
for moral improvement. If a real journey, perhaps from London to
Cambridge. **79 Findor** (I am split) either between two places, where
he saw her and where he now is, or between his body, which is here,
and his soul, which looks for his love. **81 proles** (offspring) Vulcan,
thrown to earth by Jove. M reworked this image in his famous lines on
Satan and Mulciber, *PL* 1 44, 740. **84 Amphiaraus** Greek prince
and seer, fleeing after defeat in the war of the Seven against Thebes,
was saved from disgrace by Jove, who caused the ground to open before
his feet. Image perhaps also recalls death of Hippolytus, in Ovid *Ibis* 578.

But onwards I journeyed, weeping in silence, deranged,
confused, and often want to return there again.
I am split: half here, while the rest of me looks for my
love.
When joys suddenly vanish, weeping is sweet.   80
So Juno's son lamented the heaven he'd lost,
hurled headlong down to Lemnos' burning hearths.
Thus Amphiaráus looked back to the sun he was torn from
when frenzied horses carried him down to hell.

I'm desperate, conquered by grief, but what can I do?   85
Neither continue this love, nor lay it aside.
If only a god would grant that I look on those features
I love, and speak sad words to her, face to face.
Perhaps she was not created from hardest adamant,
maybe she'd not be deaf to all my prayers.   90
Surely no-one has burned in so luckless a love!
I pray that mine is your first and only such case.
Spare me I beg you, O Love; you are tender, and winged,
signs of your function your deeds should never oppose.
For you know that now I dread the might of your bow,   95
you son of a goddess, your darts as potent as fire.
And I'll make your holy altars smoke with my offerings,
and call you the only chief of the gods above.
But finally take my – no, leave me my madness, I mean:
somehow, pain feels sweet to all who love.   100
But still, be mild with your gifts: if I love again,
let the love be mutual, one arrow piercing us both.

91 nullus (no-one) either he sees himself as Apollo, Cupid's first
exemplary victim, or is the unhappiest of all, and hence will serve as
Cupid's only example of his power. **infeliciter** (unsuccessful, or unhappy). **93 ales** (winged) symbol of love's rapidity of flight, changeability. M wants to switch affections quickly. **98 Solus** (alone). **summus**
(highest) these compete with each other, suggesting perhaps his in-
sincerity. **102 una** (one) juxtaposed to **duos**, two. Cupid had had
two arrows for Apollo and Daphne, one gold-tipped to make Apollo
love, the other lead-tipped to make Daphne flee.

[*Epilogue to the elegies*]

Haec ego mente olim laeva, studioque supino
    Nequitiae posui vana trophaea meae.
    Scilicet abreptum sic me malus impulit error,
Indocilisque aetas prava magistra fuit.
Donec Socraticos umbrosa Academia rivos   5
    Praebuit, admissum dedocuitque iugum.
Protinus extinctis ex illo tempore flammis,
    Cincta rigent multo pectora nostra gelu.
Unde suis frigus metuit puer ipse sagittis,
    Et Diomedeam vim timet ipsa Venus.   10

1 **laeva** (left) of the many connotations of left, those relevant here include ill-omened, foolish; also the heart is on the left. **supino** lit. lying on one's back; hence idle, retrogressive. A paradoxical conjunction of qualities, slack energy. 2 **Nequitiae** (vileness, debauchery) sins of weakness of character, self-indulgence rather than the positive evil

[*Epilogue to the elegies*]

IN THE PAST, with part of my mind, and perverse energy
    I composed these idle trophies to celebrate vice.
  I was fallen from grace, and driven by error and sin,
    my corrupt master was youth, which refused to be
      taught:
till Plato's shady academy offered me Socrates'   5
    streams, and I learnt to unlearn what enslaved me
      before.
At once my flames were extinguished, and still they are
      dead:
  and my breast is frozen, bound with layers of ice.
So Cupid himself is afraid for the cold on his arrows,
    and Venus fears my strength, Diomédes returned.   10

suggested in the next line. **3 Scilicet** (certainly) either intensive, or
perhaps protesting too much? **5 Socraticos** (Socratic) Plato, pupil of
Socrates, who taught that one should pass from earthly loves to love of
the idea of the good. **10 Diomedeam** Diomedes wounded Venus when
she intervened in battle to protect her son Aeneas (*Iliad* v 334).

The fifth ode of Horace, Book I

Quis multa gracilis te puer in rosa

*Rendered almost word for word without rhyme according to
the Latin measure, as near as the language will permit*

WHAT SLENDER YOUTH bedewed with liquid odours
  Courts thee on roses in some pleasant cave,
   Pyrrha for whom bind'st thou
   In wreaths thy golden hair,
Plain in thy neatness; O how oft shall he   5
On faith and changed gods complain: and seas
   Rough with black winds and storms
   Unwonted shall admire:
Who now enjoys thee credulous, all gold,
Who always vacant always amiable   10
   Hopes thee; of flattering gales
   Unmindful? Hapless they
To whom thou untried seem'st fair. Me in my vowed
Picture the sacred wall declares to have hung
   My dank and dropping weeds   15
   To the stern god of sea.

8 Unwonted he will be unused to it. **admire** close to Lat *emirabitur* he
will be amazed. **9 credulous** this is the youth. M is using the free word
order of the Latin. Did he mean to suggest that she is credulous too?
10 vacant again using Lat word (*vacuam* unengaged, without other lovers)
and word order. **amiable** Lat *amabilem* loving. **11 Hopes** Lat construction ('hopes you will be'). **gales** winds. **13 fair** for Lat *nites* (you
shine, glitter), appropriate to either girl or sea. **13–16** M's renunciation
of the storm of sexuality. His syntax follows the Lat very closely.
vowed during a storm a sailor might promise to place a plaque in the
temple of a god who saved him. **15 dropping** for dripping? **weeds**
clothes.

Song. *On May morning*

NOW THE BRIGHT MORNING STAR, day's harbinger,
Comes dancing from the east, and leads with her
The flowery May, who from her green lap throws
The yellow cowslip, and the pale primrose.
Hail bounteous May that dost inspire   5
Mirth and youth and warm desire,
Woods and groves are of thy dressing,
Hill and dale doth boast thy blessing.
Thus we salute thee with our early song,
And welcome thee, and wish thee long.   10

1 **star** for Venus. 6 **youth** to inspire youth means to make young.
Cf *El V* etc. 7 **groves** if there's a difference between woods and groves
it would be that groves are smaller and more cultivated.

Sonnet 1 *O nightingale*

O NIGHTINGALE, that on yon bloomy spray
    Warblest at eve, when all the woods are still,
    Thou with fresh hope the lover's heart dost fill,
While the jolly hours lead on propitious May,
Thy liquid notes that close the eye of day,   5
    First heard before the shallow cuckoo's bill
    Portend success in love; O if Jove's will
    Have linked that amorous power to thy soft lay,
Now timely sing, ere the rude bird of hate
    Foretell my hopeless doom in some grove nigh: 10
    As thou from year to year hast sung too late
For my relief; yet hadst no reason why,
    Whether the Muse, or Love call thee his mate,
    Both them I serve, and of their train am I.

1 nightingale associated by M elsewhere with melancholy, evening (*PL* v 40) and a plangent kind of love ('amorous descant' *PL* IV 602). **bloomy** with flowers on, suggesting spring, success in love. This first quatrain is optimistic. **5 liquid** clear, flowing (like Latin *liquidus*). **6–7** refers to a tradition, which M thought came from Chaucer, that to hear the nightingale first was a good omen for a lover. M gives the cuckoo-qualities as shallow (fickle) rude (mere lust, not the gentle love of the nightingale) hate (implacable hostility towards the lover; a hard mistress). So the opposition between nightingale and cuckoo is not between love and no love, but between two kinds of love. **9 timely** at the right time, ie now. This shows that she isn't singing now, in spite of 'warblest' of line 2, so the apparent reality of the first 8 lines dissolves into general propositions. **12–14** difficult sense. The nightingale (and hence M too) has no reason why M serves both the Muse (poetry, Apollo) and Love (Cupid). M doesn't know which of these the nightingale serves, either. Some editors repunctuate with comma instead of semi-colon after *relief*, and a colon after *why*.

Sonnet 2   *Donna leggiadra*

DONNA LEGGIADRA il cui bel nome honora
L'herbosa val di Rheno, e il nobil varco,
Bene è colui d'ogni valore scarco
Qual tuo spirto gentil non innamora,
Che dolcemente mostra si di fuora   5
De sui atti soavi giamai parco,
E i don', che son d'amor saette ed arco,
Là onde l' alta tua virtù s'infiora.
Quando tu vaga parli, o lieta canti
Che mover possa duro alpestre legno,   10
Guardi ciascun a gli occhi, ed a gli orecchi
L'entrata, chi di te si truova indegno;
Gratia sola di sù gli vaglia, inanti
Che'l disio amoroso al cuor s'invecchi.

Sonnet 3   *Qual in colle aspro*

QUAL IN COLLE ASPRO, al imbrunir di sera
L'avezza giovinetta pastorella
Va bagnando l'herbetta strana e bella
Che mal si spande a disusata spera
Fuor di sua natia alma primavera,   5
Cosi amor meco insù la lingua snella
Desta il fior novo di strania favella,
Mentre io di te, vezzosamente altera,
Canto, dal mio buon popol non inteso
E'l bel Tamigi cangio col bel Arno.   10

[*Sonnet 2*]   **4 Qual** whom, object of the verb (the lover is passive before this beauty). **7 don'** (gifts) of God to her, and her to everyone else? **8 Là** (there) where? Probably her eyes. **s'infiora** (enflowers itself) see Dante's *Par* XXXI 7, of bees/angels immersing themselves in the white rose of the heavenly host.   **10** suggests myth of Orpheus, who could charm even stocks and stones with his song. He represents the power of poetry, not usually virtue.   **14 s'invecchi** literally 'in-olds itself', the form reflecting *s'infiora* in 8. The idea is that once desire becomes habitual it is impossible to remove. [*Sonnet 3*]   **1 Qual** (as) beginning of long suspended

Sonnet 2

BEAUTIFUL LADY, whose name adds honour
to Reno's grassy valley and famous ford,
truly the man is void of every virtue
who fails to fall in love with your noble soul,
which shows itself so sweetly abroad   5
(in unfailing bounty of tender looks,
and the favours Love employs as his bow and arrows)
there where your highest virtue is in full bloom.
Whenever you sweetly speak, or sing with joy
a song that could move the hardest pine,   10
then each must look to the portals of his eyes
and ears, whoever is found unworthy of you:
only grace from above can save him, before
desire grows old in the heart, and can't be removed.

Sonnet 3

ON A BARREN HILLSIDE as evening darkens
a little girl who lives there, tending sheep,
waters a plant that is small and strange and
    lovely
and weakly spreads its leaves to the alien sky,
far from the generous spring-time which gave it birth: 5
so also Love is waking the fresh sprung flower
of a foreign language that grows on my tripping tongue
as I sing of you, so pretty in your pride,
while my good countrymen fail to understand me,
as I change from the lovely Thames to lovely Arno.   10

construction. 2–3 note all the prettifying diminutives, *etta* and *ella*.
5 **primavera** (spring) refers to a time, balancing place in the previous
line (*spera*) but talked of as if it were spatial (*fuor di* – outside) and
a kind of soil (*alma* – benign, nourishing). 10 **Tamigi** (Thames) English
in general, and perhaps the English spoken in London particularly,
which would become the standard language, as the Tuscan dialect of
Florence, on the Arno, established itself as the proper form for literary
expression in Italian, after the literary achievements of Dante and Petrarch.

Amor lo volse, ed io a l'altrui peso
Seppi ch' Amor cosa mai volse indarno.
Deh! foss' il mio cuor lento e'l duro seno
A chi pianta dal ciel si buon terreno.

Canzone

R IDONSI DONNE E GIOVANI amorosi
   M' accostandosi attorno, e 'perche scrivi,
   Perche tu scrivi in lingua ignota e strana
Verseggiando d'amor, e come t'osi?
Dinne, se la tua speme sia mai vana,   5
E de pensieri lo miglior t' arrivi;'
Cosi mi van burlando, 'altri rivi
Altri lidi t' aspettan, ed altre onde
Nelle cui verdi sponde
Spuntati ad hor, ad hor a la tua chioma   10
L'immortal guiderdon d'eterne frondi,
Perche alle spalle tue soverchia soma?'
   Canzon dirotti, e tu per me rispondi
Dice mia Donna, e'l suo dir, è il mio cuore
'Questa è lingua di cui si vanta Amore.'   15

Sonnet 4   Diodati, e te'l dirò

D IODATI, E TE'L DIRÒ con maraviglia,
   Quel ritroso io ch'amor spreggiar soléa
   E de suoi lacci spesso mi ridéa
Gia caddi, ov'huom dabben talhor s'impiglia.

11–12 M learning by others' examples, in contrast with *El VII*.
13–14 cf parable of the sower, *Matt* xiii. [*Canzone*] 1 **Ridonsi**
(laugh) either with delight or contempt; ambiguous social relationship. *si* on the end of the word is a pseudo reflexive, giving a
more intense, perhaps more private, quality to the feelings. See also
accostandosi (crowd round) in 2, *t'osi* (dare you yourself) at 4. 5 indirect
construction; a compressed prayer or wish. **6 pensieri** means something halfway between a thought and a wish or desire, in the language
of the sonnets. **7 rivi** (streams) other languages: English and/or Latin.
10 ad hor, ad hor (now and now) either suggesting its gradual growth, or

Love decreed it, and I from others pains
know that Love has never decreed in vain.
If only my sluggish heart and stony breast
might prove as rich a soil for the sower from heaven.

Canzone

THEY LAUGH, THE LOVERS, ladies and youths
    crowding around me: 'Why do you write,
    Why write in a language remote and obscure
rhyming of love? How dare you?
Tell us: we'll pray your hopes will never be vain,   5
that things fall out as you wanted most.'
And thus they continue to mock me: 'Other rivers
and shores await you, and other waters
on whose green banks
minute by minute sprout the eternal leaves,   10
an immortal garland to crown your hair.
Why add this other intolerable weight to your back?'
    Canzone, I'll tell you, and you can answer for me.
My mistress declares – and her words are my heart –
'This is the language Love is proudest of.'   15

Sonnet 4

DIODATI – I'M AMAZED I'm telling you this –
    the bashful me, who used to mock at love
    and often laughed at the snares he laid,
I've fallen, where sometimes an upright man is caught,

insisting that this is happening this very minute. **11 guiderdon**
(reward) cf 'that immortal garland' which 'is to be run for, not without
dust and heat', M's famous image in *Areopagitica*. **12 soverchia**
(excessive) main meaning the Italian language, which M was in fact
finding difficult to master, in spite of *Sonnet 3*'s claims, but could also
refer to the burden of his body and desires. **14** the language here is
simple but it's hard to see exactly what it means. How can her speech
be his heart? **15 si vanta** (boasts) a word with critical implications,
where something more positive, like 'delights', would be expected.

Ne treccie d'oro, ne guancia vermiglia   5
  M' abbaglian sì, ma sotto nova idea
  Pellegrina bellezza che'l cuor bea,
  Portamenti alti honesti, e nelle ciglia
Quel sereno fulgor d' amabil nero,
  Parole adorne di lingua piu d'una,   10
  E'l cantar che di mezzo l'hemispero
Traviar ben può la faticosa Luna,
  E degli occhi suoi auventa si gran fuoco
  Che l'incerar gli orecchi mi fia poco.

Sonnet 5 *Per certo*

PER CERTO I BEI VOSTR' OCCHI, Donna mia
    Esser non puo che non fian lo mio sole
    Si mi percuoton forte, come ei suole
  Per l'arene di Libia chi s'invia,
Mentre un caldo vapor (ne senti pria)   5
  Da quel lato si spinge ove mi duole,
  Che forse amanti nelle lor parole
  Chiaman sospir; io non so che si sia:
Parte rinchiusa, e turbida si cela
  Scosso mi il petto, e poi n'uscendo poco   10
  Quivi d' attorno o s'agghiaccia, o s'ingiela;
Ma quanto a gli occhi giunge a trovar loco
  Tutte le notti a me suol far piovose
  Finche mia Alba rivien colma di rose.

[*Sonnet 4*] **5 treccie** (tresses) rejection of the traditional golden-haired
beauty is itself part of the convention; see eg Sidney and Shakespeare.
6 idea (model, pattern) reference to neo-platonic conceptions, whereby
various beautiful ladies embody in different ways the single idea of
beauty and goodness. It's heretical for a neo-platonist to talk of a new
idea/pattern, because the patterns were eternal, and only embodiments
could be new. **9 nero** (black) either her eyebrows or eyelashes, or the

No tresses of gold or rosy cheeks   5
   have bewildered me so, but a foreign beauty
   and new pattern of loveliness delights my heart;
   modestly proud in her bearing, and in her eyes
what a lovely blackness glowing serenely bright;
   a fluent linguist, accomplished in several tongues: 10
   and able with just her singing to drive the moon
as it busily rides on high astray from its course:
   and forth from her eyes she shoots so fierce a flame
   that wax in my ears would do me little good.

Sonnet 5

SURELY YOUR LOVELY EYES, my lady,
   cannot be anything else except my sun;
   they beat upon me as fiercely as does the sun
on someone travelling across the Libyan sands,
while a scalding gas (not something I'd felt before)   5
   spurts on that side which feels most pain –
   perhaps in the language lovers use
it's called a sigh: I don't know what it could be
and part is compressed, and swirls in hiding
   and shakes my breast: and when a little escapes   10
   to the outside air it freezes or turns to ice.
But as much as manages to find its way to my eyes
   is likely to fill my every night with rain
   till my Dawn returns, brimming with roses.

eyes that nestle within them. **12 faticosa** either industrious, so that
the (feminine) moon on this occasion is like M, or 'in eclipse', cf *PL* ii
665. **14** Ulysses put wax in the ears of his sailors, to stop them hearing
the dangerous song of the sirens. The image suggests that the beauty
of this girl might not be an entirely good influence. [*Sonnet 5*] **6
lato** (side) the left side, where the heart is. **11 s'agghiacci** (turns to
ice) in the absence of the sun/mistress.

At a vacation exercise in the college

The Latin speeches ended, the English thus began

HAIL NATIVE LANGUAGE, that by sinews weak
Didst move my first endeavouring tongue to speak,
And mad'st imperfect words with childish trips,
Half unpronounced, slide through my infant lips,
Driving dumb silence from the portal door,　　　　5
Where he had mutely sat two years before:
Here I salute thee and thy pardon ask,
That now I use thee in my latter task:
Small loss it is that thence can come unto thee,
I know my tongue but little grace can do thee.　　　10
Thou need'st not be ambitious to be first,
Believe me I have thither packed the worst:
And, if it happen as I did forecast,
The daintiest dishes shall be served up last.
I pray thee then deny me not thy aid　　　　15
For this same small neglect that I have made;
But haste thee straight to do me once a pleasure,
And from thy wardrobe bring thy chiefest treasure;
Not those new-fangled toys, and trimming slight
Which takes our late fantastics with delight,　　　20
But cull those richest robes, and gayest attire
Which deepest spirits, and choicest wits desire:
I have some naked thoughts that rove about
And loudly knock to have their passage out;

1 by sinews weak goes either with *move*, in which case language was
as weak as the infant M, or with *speak*, where it would be only M who
was weak. **2 first endeavouring** departing drastically from natural
English word-order, compressing a clause 'when my tongue first
endeavoured to speak'. **4 slide** M not seemingly responsible for his
speech, for wanting to say something. **6 mutely** didn't the baby M
cry? **8 latter** the first part of the exercise was in Latin prose. **16
neglect** to *make* a neglect is odd for English; turning a lack of action
into a product. **20 late** recent. **fantastics** people who use their fantasy
or imagination. **22 deepest** probably has the ambiguity of the Latin
altus, meaning high as well as deep (see 33). **choicest** most select – who
is doing the selecting?

And weary of their place do only stay 25
Till thou hast decked them in thy best array;
That so they may without suspect or fears
Fly swiftly to this fair assembly's ears;
Yet I had rather, if I were to choose,
Thy service in some graver subject use, 30
Such as may make thee search thy coffers round,
Before thou clothe my fancy in fit sound:
Such where the deep transported mind may soar
Above the wheeling poles, and at heaven's door
Look in, and see each blissful deity 35
How he before the thunderous throne doth lie,
Listening to what unshorn Apollo sings
To the touch of golden wires, while Hebe brings
Immortal nectar to her kingly sire:
Then passing through the spheres of watchful fire, 40
And misty regions of wide air next under,
And hills of snow and lofts of pilèd thunder,
May tell at length how green-eyed Neptune raves,
In heaven's defiance mustering all his waves;
Then sing of secret things that came to pass 45
When beldam Nature in her cradle was;
And last of kings and queens and heroes old,
Such as the wise Demodocus once told
In solemn songs at king Alcinous' feast,
While sad Ulysses' soul and all the rest 50
Are held with his melodious harmony

27 suspect suspicion. Why should his ideas be fearful or suspicious?
32 fancy imagination – so he is a 'fantastic' too. **33 deep** as in
22, both deep and high (hence 'soar'), from Latin *altus* – M using
the potentialities of Latin to enrich his English even when he is
supposedly greeting his native language. **36 he** but there were
female deities too. **37 unshorn** unlike Samson in Gaza. **38 Hebe**
goddess of youth. **40 watchful fire**, spheres of the Ptolemaic universe, each made of aetherial fire, and also living, able to look down
on earth below (M using a poetic-platonic rather than scientific version
here). **43 raves** another heavenly rebellion envisaged by the youthful
M. **46 beldam** lovely lady, but a homely phrase, suggesting a cosmic
Granny. **48 Demodocus** poet at the court of the virtuous King
Alcinous, where Ulysses rested on his journey home (*Odyssey* 8).

In willing chains and sweet captivity.
But fie my wandering Muse how thou dost stray!
Expectance calls thee now another way,
Thou knowest it must be now thy only bent   55
To keep in compass of thy predicament:
Then quick about thy purposed business come,
That to the next I may resign my room.

56 compass narrow circle. **predicament** pun, both 'difficulty', and 'category' in Aristotelian logic.

Ad Joannem Rousium Oxoniensis Academiae Bibliothecarium

Strophe i

GEMELLE CULTU SIMPLICI gaudens liber,
    Fronde licet gemina,
    Munditieque nitens non operosa,
Quam manus attulit
Iuvenilis olim,   5
Sedula tamen haud nimii poetae;
Dum vagus Ausonias nunc per umbras
Nunc Britannica per vireta lusit
Insons populi, barbitoque devius
Indulsit patrio, mox itidem pectine Daunio   10
Longinquum intonuit melos
Vicinis, et humum vix tetigit pede...

(Antistrophe i not included]

Strophe ii

Modo quis deus, aut editus deo   25
Pristinam gentis miseratus indolem
(Si satis noxas luimus priores
Mollique luxu degener otium)
Tollat nefandos civium tumultus,
Almaque revocet studia sanctus   30
Et relegatas sine sede musas
Iam pene totis finibus Angligenum;
Immundasque volucres
Unguibus imminentes
Figat Apollinea pharetra,   35
Phineamque abigat pestem procul amne Pegaseo.

1 **Gemelle** (twin) identical to the book that was stolen. **cultu simplici** a paradox, made more complex by the ambiguity of *cultu*, either adornment, which isn't usually simple, or dress, the cover, which might have been in unostentatious leather. 2 **Fronde** (bough, garland) perhaps a double garland for the two languages, English and Latin, cf *Canzone* 7–12, or perhaps the double frontispiece, since M had the two sets of verse bound together. 7 **umbras** (shades) probably refers to poetry written before his trip to the continent, so a journey becomes a metaphor

To John Rouse, librarian of the University of Oxford

Strophe i

THIS BOOK IS A TWIN, that delights in its simple
but sprouts a double leaf,   [elegance,
shining with casual polish
which a youthful hand
once brought it,   5
earnest, but not quite a poet yet:
who wandered in shady groves in Italy
or played in Britain's green glades
offending no-one, indulging in secret
his native lyre, then singing a foreign strain   10
in Italian modes
to his neighbours at home, feet hardly touching the ground...

[Antistrophe i not included]

Strophe ii

What god or child of a god   25
will pity our race for its native virtues
(if we've atoned enough for our former sins,
degenerate sloth and slack debauch)
and end these criminal riots of civil war:
some holy one, to bring back fruitful studies   30
and the Muses, banished now with scarcely
a resting place in all the English realm:
to shoot with Apollo's arrows
the filthy Harpies
whose claws are poised to strike,   35
and to drive the plague of Phineas far from Pegasus' streams.

for use of the language. Or perhaps *umbra* has its derived meaning of
images. **25 editus** (begotten by) especially used of books, so there is
probably a pun here. **26 indolem** (native character or virtues) usually
in a positive sense, but in his *History of Britain* M was unable to find
a time when the British had ever shown admirable qualities. **36 pestem**
(plague) the Harpies, winged creatures sent by Jove to snatch food from
blind Phineas. Sometimes allegorized as mean-minded critics.

On the death of a fair infant dying of a cough

I

O FAIREST FLOWER no sooner blown but blasted,
    Soft silken primrose fading timelessly,
    Summer's chief honour if thou hadst outlasted
Bleak winter's force that made thy blossom dry;
For he being amorous on that lovely dye   5
    That did thy cheek envermeil, thought to kiss
But killed alas, and then bewailed his fatal bliss.

II

For since grim Aquilo his charioteer
By boisterous rape the Athenian damsel got,
He thought it touched his deity full near,   10
If likewise he some fair one wedded not,
Thereby to wipe away the infamous blot
    Of long-uncoupled bed, and childless eld,
Which 'mongst the wanton gods a foul reproach was held.

III

So mounting up in icy-pearled car,   15
Through middle empire of the freezing air
He wandered long, till thee he spied from far,
There ended was his quest, there ceased his care.
Down he descended from his snow-soft chair,
    But all unwares with his cold-kind embrace   20
Unhoused thy virgin soul from her fair biding-place.

2 primrose perhaps the Latin *prima rosa*, first rose, since the primrose
is yellow but this flower has a pink cheek in 6. **timelessly** outside
time? For ever? At the wrong time, too early, would be the obvious
meaning. **4 dry** more usual effect of summer than of winter. **6
envermeil** make vermilion. **8 Aquilo** the North wind, who carried
off Orithyia, an Athenian princess. **13 eld** old age, an archaic form.
16 middle empire region above the sphere of the moon. Cf *Vac Ex* 41,
written not long after this, where M is rather like Winter in this poem.

IV

Yet art thou not inglorious in thy fate;
For so Apollo, with unweeting hand
Whilom did slay his dearly-loved mate
Young Hyacinth born on Eurotas' strand   25
Young Hyacinth the pride of Spartan land;
   But then transformed him to a purple flower
Alack that so to change thee winter had no power.

V

Yet can I not persuade me thou art dead
Or that thy corse corrupts in earth's dark womb,
Or that thy beauties lie in wormy bed,
Hid from the world in a low-delved tomb;
Could heaven for pity thee so strictly doom?
   O no! for something in thy face did shine
Above mortality that showed thou wast divine.   35

VI

Resolve me then O soul most surely blest
(If so it be that thou these plaints dost hear)
Tell me bright spirit where'er thou hoverest
Whether above that high first-moving sphere
Or in the Elysian fields (if such there were)   40
   O say me true if thou wert mortal wight
And why from us so quickly thou didst take thy flight.

VII

Wert thou some star which from the ruined roof
Of shaked Olympus by mischance didst fall;
Which careful Jove in nature's true behoof   45

23 **unweeting** hand which did not know what it was doing. Apollo
threw a discus, which accidentally killed Hyacinth, his boyfriend. **27**
transformed as M has transformed Anne to a flower. **30 womb**
linked by assonance to *wormy* and by rhyme to *tomb*: birth and death.
32 delved dug, archaic form. **low** here could be either deep or shallow.
40 were tense here implies that they are not. **41 wight** person (archaic).
45 true behoof Nature's real good, in contrast to her apparent needs.
Jove here seems very domestic, but who or what shook Olympus?

Took up, and in fit place did reinstall?
Or did of late Earth's sons besiege the wall
  Of sheeny heaven, and thou some goddess fled
Amongst us here below to hide thy nectared head?

VIII

Or wert thou that just maid who once before   50
Forsook the hated earth, O tell me sooth
And cam'st again to visit us once more?
Or wert thou that sweet smiling youth Mercy?
Or that crowned matron sage white-robed Truth?
  Or any other of that heavenly brood   55
Let down in cloudy throne to do the world some good?

IX

Or wert thou of the golden-winged host,
Who having clad thyself in human weed,
To earth from thy prefixed seat didst post,
And after short abode fly back with speed,   60
As if to show what creatures heaven doth breed,
  Thereby to set the hearts of men on fire
To scorn the sordid world, and unto heaven aspire?

X

But O why didst thou not stay here below
To bless us with thy heaven-loved innocence,   65
To slake his wrath whom sin hath made our foe
To turn swift-rushing black perdition hence,
Or drive away the slaughtering pestilence,
  To stand 'twixt us and our deserved smart?
But thou canst best perform that office where thou art.   70

47 **Earth's sons** the giants, sons of Earth, tried to overthrow Jove.
50 **just maid** Astraea, last of the gods to leave the earth in disgust
at the crimes of men, her return was to be the sign of a new golden
age. 53 **Mercy** here is an editorial guess. Without it, M's line is a foot
short and difficult to interpret. In any case, what is the sex of the
youth? 59 **prefixed** predestined. 67 **perdition** perhaps a specific
plague, though there was no outbreak at the time of composition.
70 **office** she sounds here like the virgin Mary in catholic doctrine.

XI

Then thou the mother of so sweet a child
Her false imagined loss cease to lament,
And wisely learn to curb thy sorrows wild;
Think what a present thou to God hast sent,
And render him with patience what he lent;   75
   This if thou do he will an offspring give,
That till the world's last end shall make thy name to live.

73 **wild** the insane grief suggested here hasn't been registered strongly
in the rest of the poem.   77 **last** redundant since ends usually are last.

On the University Carrier

who sickened in the time of his vacancy, being forbid to go to London, by reason of the plague

HERE LIES OLD HOBSON, Death hath broke his girt,
  And here alas, hath laid him in the dirt,
  Or else the ways being foul, twenty to one,
He's here stuck in a slough, and overthrown.
'Twas such a shifter, that if truth were known,   5
Death was half glad when he had got him down;
For he had any time this ten years full,
Dodged with him, betwixt Cambridge and the Bull.
And surely, Death could never have prevailed,
Had not his weekly course of carriage failed;   10
But lately finding him so long at home,
And thinking now his journey's end was come,
And that he had ta'en up his latest inn,
In the kind office of a chamberlain
Showed him his room where he must lodge that night,   15
Pulled off his boots, and took away the light:
If any ask for him, it shall be said,
Hobson has supped, and 's newly gone to bed.

Another on the same

HERE LIETH ONE who did most truly prove,
  That he could never die while he could move,
  So hung his destiny never to rot
While he might still jog on and keep his trot,
Made of sphere-metal, never to decay   5
Until his revolution was at stay.

---

1 girt the girth of his saddle.   **4 slough** miry ditch or muddy patch in the road.   **5 shifter** trickster, good at dodging.   **8 the Bull** London inn where Hobson's journey finished.   **14 chamberlain** servant in an inn who attended the bedrooms.   **5 sphere-metal** the spheres were made of an imperishable substance.   **6 at stay** had stopped.

Time numbers motion, yet (without a crime
'Gainst old truth) motion numbered out his time:
And like an engine moved with wheel and weight,
His principles being ceased, he ended straight,   10
Rest that gives all men life, gave him his death,
And too much breathing put him out of breath;
Nor were it contradiction to affirm
Too long vacation hastened on his term.
Merely to drive the time away he sickened,   15
Fainted, and died, nor would with ale be quickened,
Nay, quoth he, on his swooning bed outstretched,
If I may not carry, sure I'll ne'er be fetched,
But vow though the cross doctors all stood hearers,
For one carrier put down to make six bearers.   20
Ease was his chief disease, and to judge right,
He died for heaviness that his cart went light,
His leisure told him that his time was come,
And lack of load, made his life burdensome,
That even to his last breath (there be that say't)   25
As he were pressed to death, he cried more weight;
But had his doings lasted as they were,
He had been an immortal carrier.
Obedient to the moon he spent his date
In course reciprocal, and had his fate   30
Linked to the mutual flowing of the seas,
Yet (strange to think) his wain was his increase:
His letters are delivered all and gone,
Only remains this superscription.

7 motion depends on time so that when his motion was zero he had no
more time. 10 principles double sense: mechanically it's the initiator of
movement, philosophically the principles he stood for. 19 cross
antagonistic. 20 bearers pall-bearers. 26 criminals were pressed to
death by weights on their chest, to induce them to confess their crime.
30 reciprocal corresponding to the moon's course. 32 wain waggon,
also pun on wane, to continue link with the moon.

An epitaph on the Marchioness of Winchester

THIS RICH MARBLE doth inter
The honoured wife of Winchester,
A viscount's daughter, an earl's heir,
Besides what her virtues fair
Added to her noble birth,   5
More than she could own from earth.

Summers three times eight save one
She had told, alas too soon,
After so short time of breath,
To house with darkness, and with death.   10
Yet had the number of her days
Been as complete as was her praise,
Nature and fate had had no strife
In giving limit to her life.
Her high birth, and her graces sweet,   15
Quickly found a lover meet;
The virgin choir for her request
The god that sits at marriage-feast;
He at their invoking came
But with a scarce-well-lighted flame;   20
And in his garland as he stood,
Ye might discern a cypress bud.
Once had the early matrons run
To greet her of a lovely son,
And now with second hope she goes,   25
And calls Lucina to her throes;
But whether by mischance or blame
Atropos for Lucina came;

17 **virgin choir** actually it would have been her father and the father
of the chosen husband doing the requesting.   18 Hymen.   22 **cypress**
symbol of death.   23 **early matrons run** for 'the matrons (midwives?)
had come before the birth', a strained latinate construction. M retreating
in embarrassment from the scene of birth?   26 **Lucina** Roman goddess of child-birth.   27 **blame** guilt – of the virtuous Jane?   28
Atropos one of the Fates, whose task was to cut the thread of life.

And with remorseless cruelty,
Spoiled at once both fruit and tree: 30
The hapless babe before his birth
Had burial, yet not laid in earth,
And the languished mother's womb
Was not long a living tomb.

So have I seen some tender slip   35
Saved with care from winter's nip,
The pride of her carnation train,
Plucked up by some unheedy swain,
Who only thought to crop the flower
New shot up from vernal shower;   40
But the fair blossom hangs the head
Sideways as on a dying bed,
And those pearls of dew she wears,
Prove to be presaging tears
Which the sad morn had let fall   45
On her hastening funeral.

Gentle lady may thy grave
Peace and quiet ever have;
After this thy travail sore
Sweet rest seize thee evermore,   50
That to give the world increase,
Shortened hast thy own life's lease;
Here, besides the sorrowing
That thy noble house doth bring,
Here be tears of perfect moan   55
Wept for thee in Helicon,
And some flowers, and some bays,
For thy hearse to strew the ways,
Sent thee from the banks of Came,

32 **yet** still. 33 **languished** who had languished and died. **37 carnation** pink, flesh-coloured. **40 vernal** spring. **shower** up–down juxtaposition with flower shooting **up**. **50 seize** the English word sounds too strong for *rest*. M probably had the Latin *capere* in mind. **59 Came** the river Cam, for Cambridge University.

Devoted to thy virtuous name;   60
Whilst thou bright saint high sit'st in glory,
Next her much like to thee in story,
That fair Syrian shepherdess,
Who after years of barrenness,
The highly favoured Joseph bore   65
To him that served for her before,
And at her next birth much like thee,
Through pangs fled to felicity,
Far within the bosom bright
Of blazing majesty and light,   70
There with thee, new welcome saint,
Like fortunes may her soul acquaint,
With thee there clad in radiant sheen,
No marchioness, but now a queen.

63 Rachel, in fact Jacob's second wife, since her father made Jacob
marry her elder sister Leah first. Leah bore Jacob 6 sons while Rachel
remained barren, and Rachel was reduced to proxy birth, her maid
Bilhah giving birth to sons by Jacob on Rachel's knees. Rachel gave
birth to Joseph, then died after giving birth to Benjamin (*Gen* xxix 35).

On Shakespeare

WHAT NEEDS MY SHAKESPEARE for his
      honoured bones,
        The labour of an age in piled stones,
Or that his hallowed relics should be hid
Under a star-ypointing pyramid?
Dear son of memory, great heir of fame,   5
What need'st thou such weak witness of thy name?
Thou in our wonder and astonishment
Hast built thyself a live-long monument.
For whilst to the shame of slow-endeavouring art,
Thy easy numbers flow, and that each heart   10
Hath from the leaves of thy unvalued book,
Those Delphic lines with deep impression took,
Then thou our fancy of itself bereaving,
Dost make us marble with too much conceiving;
And so sepulchred in such pomp dost lie,   15
That kings for such a tomb would wish to die.

1 my asserts M's intimacy with Shakespeare. **4 ypointing** meant to
sound archaic. **5 son, heir** Shakespeare without noble ancestors, like
the Marchioness's. Since his fame and the memory of him lives after
not before him, inheritance seems to work in reverse for him. **9 slow-
endeavouring** Shakespeare was often contrasted especially with Ben
Jonson for the speed with which he composed (cf *L'All* 132–4). M would
probably have seen himself as more like Jonson in this respect.
11 unvalued invaluable? or, not valued at first. **12 Delphic** of the
Delphic oracle, the shrine of Apollo.

Epitaph for Damon

Argument [Latin not printed here]. Thyrsis and Damon, shepherds
from the same neighbourhood, had pursued the same interests
since boyhood, and been the closest of friends. Thyrsis, travelling
abroad to improve his mind, received news of Damon's death.
Later, having returned home and found that this news was true,

Epitaphium Damonis

HIMERIDES NYMPHAE (nam vos et Daphnin et
    Hylan,
    Et plorata diu meministis fata Bionis)
Dicite Sicelicum Thamesina per oppida carmen:
Quas miser effudit voces, quae murmura Thyrsis,
Et quibus assiduis exercuit antra querelis,   5
Fluminaque, fontesque vagos, nemorumque recessus,
Dum sibi praereptum queritur Damona, neque altam
Luctibus exemit noctem loca sola pererrans.

Et iam bis viridi surgebat culmus arista,
Et totidem flavas numerabant horrea messes,   10
Ex quo summa dies tulerat Damona sub umbras,
Nec dum aderat Thyrsis; pastorem scilicet illum
Dulcis amor musae Thusca retinebat in urbe.
Ast ubi mens expleta domum, pecorisque relicti
Cura vocat, simul assueta seditque sub ulmo,   15
Tum vero amissum tum denique sentit amicum,
Coepit et immensum sic exonerare dolorem.
   Ite domum impasti, domino iam non vacat, agni.

'Hei mihi! quae terris, quae dicam numina coelo,
Postquam te immiti rapuerunt funere Damon;   20

---

1 **Himerides** (of Himera) river in Sicily. **Daphnin** shepherd boy
subject of Theocritus' first *Idyll*, Virgil *Ec* v. **Hylan** Hercules' boy
friend. One attraction of pastorals for M's present purposes was the
number of homosexual relationships it celebrated. 2 **Bionis** Bion's
Epitaph for Adonis was an important work in the development of the
pastoral genre, and he was the subject of Moschus' *Epitaph for Bion*.
5 **assiduis** (unceasing) a stock epithet, but catching M's obsessive

he mourned for himself and his loneliness in this poem. Damon
stands for Charles Diodati, who was descended on his father's
side from the Tuscan city of Lucca but was in every other way
English: in wit and learning and all other gifts he was, while he
lived, a young man of extraordinary worth.

O HIMERA'S NYMPHS, you remember Daphnis and
    Hylas
    still, and Bion's long-lamented death,
so sing a Sicilian song through towns by the Thames:
of the words and groans that Thyrsis lavished in grief,
his endlessly laboured laments which troubled the caves   5
and rivers and winding brooks and depths of groves,
mourning the premature death of his Damon, and filling
deep night with his cries, as he wandered by lonely ways.

Twice now the stalk has sprouted with unripe grain,
and twice they've counted the golden crops in their   10
    barns,
since the last day took Damon down to the shades,
and Thyrsis still was absent; no doubt sweet love
of the muse detained that shepherd in Tuscan Florence.
But his mind glutted, concern for the flocks he'd left
then called him home, and he sat by his favourite elm,   15
and then, then, then, he knew he had lost his friend,
and began to release his immense burden of grief.
   Go homewards hungry, my lambs, your master is busy.

'Alas! what gods shall I pray to, in earth or heaven,
Damon! now that they've snatched you to pitiless death. 20

industry. **7 altam** high/deep: it seems deep now, in his gloom, but
will seem high at the end of the poem, when Damon ascends. **9 viridi**
(green) unripe, like Damon. **12 scilicet** surely, of course. Perhaps
ironic here. **13** Florence. **14 expleta** (filled to the brim) either
with the sights abroad (the guilty admission that at first he was too
interested in his grand tour to register Damon's death) or over-flowing
with grief. **16 sentit** (felt, knew) either heard reliably for the first
time, or, more likely, felt inwardly, truly experienced. **18** Bion's and
Moschus' *Epitaphs* both had refrains like this. **19 dicam** either 'shall
I call on' or 'shall I call/regard as a true deity'.

Siccine nos linquis, tua sic sine nomine virtus
Ibit, et obscuris numero sociabitur umbris?
At non ille, animas virga qui dividit aurea,
Ista velit, dignumque tui te ducat in agmen,
Ignavumque procul pecus arceat omne silentum.   25
Ite domum impasti, domino iam non vacat, agni.

Quicquid erit, certe nisi me lupus ante videbit,
Indeplorato non comminuere sepulchro,
Constabitque tuus tibi honos, longumque vigebit
Inter pastores: illi tibi vota secundo   30
Solvere post Daphnin, post Daphnin dicere laudes
Gaudebunt, dum rura Pales, dum Faunus amabit:
Si quid id est, priscamque fidem coluisse, piumque,
Palladiasque artes, sociumque habuisse canorum.
Ite domum impasti, domino iam non vacat, agni.   35

Haec tibi certa manent, tibi erunt haec praemia Damon,
At mihi quid tandem fiet modo? quis mihi fidus
Haerebit lateri comes, ut tu saepe solebas
Frigoribus duris, et per loca foeta pruinis,
Aut rapido sub sole, siti morientibus herbis?   40
Sive opus in magnos fuit eminus ire leones
Aut avidos terrere lupos praesepibus altis;
Quis fando sopire diem, cantuque solebit?
Ite domum impasti, domino iam non vacat, agni.

Pectora cui credam? quis me lenire docebit   45
Mordaces curas, quis longam fallere noctem
Dulcibus alloquiis, grato cum sibilat igni

---

22 obscuris both obscure, humble, and hard to see in the shadows of hell.
23 ille (the one) probably Mercury.   **25 pecus** (herd, flock) used as a term
of contempt, it jars with the refrain, which recalls the tender concern of
the shepherd for his flock. **silentum** about whom fame is silent.   **27**
refers to a superstition that a man seen by a wolf is struck dumb. From
Virgil *Ec* ix 53.   **29 tuus tibi** (your, for you) double emphasis on Damon,
almost bitter.   **31 Daphnin** double reference to Daphnis intended
perhaps to recall Theocritus and Virgil, both of whom wrote about
Daphnis, acknowledging them as M's masters.   **34 Palladias** the arts

Will you leave me like this? Your virtue go nameless to hell,
and merge with the shades of the undistinguished dead?
But the one who sorts the souls with his golden rod
must never permit that: but lead you to join your peers,
aloof from the worthless herd of the silent dead.   25
 Go homewards hungry, my lambs, your master is busy.

But whatever should happen, unless I'm bewitched by a
  wolf,
you'll not crumble to dust in an unwept grave;
your glory will last to console you, flourishing long
among shepherds. Your name will come in their vows   30
  following
Daphnis, they'll praise you as second only to Daphnis,
for as long as Pales and Faunus cherish the land –
they will if it counts to have followed the ancient faith,
and wisdom and learning, and had a poet as friend.
 Go homewards hungry, my lambs, your master is busy.   35

These rewards are certain, and lasting, Damon;
but what will become of *me* now? What loyal friend
will stick by my side as you did time and again
when the winter was harsh, where the ground budded with
  frost;
or under a savage sun, with the parched grass dying,   40
when our task was to stalk on the trail of mighty lions,
or to frighten ravening wolves from pens on the hills?
What friend will drug my days with talking and song?
 Go homewards hungry, my lambs, your master is busy.

And who can I trust? Will anyone make me relax   45
when anxieties gnaw me? Or trick the night of its length
with sweet haranguing, as soft ripe pears hiss
in the cheerful fire, and the grate crackles with nuts, while
  outside

of Pallas Athena, goddess of wisdom and science. **39 foeta** (pregnant)
its associations clashing with those of frost. **41 leones** (lions) what
are these an analogy for? M has to stalk them 'from a distance' (**eminus**).
47 alloquiis suggested a sustained address, not desultory conversation.

Molle pyrum, et nucibus strepitat focus, at malus auster
Miscet cuncta foris, et desuper intonat ulmo.
Ite domum impasti, domino iam non vacat, agni.   50

Aut aestate, dies medio dum vertitur axe,
Cum Pan aesculea somnum capit abditus umbra,
Et repetunt sub aquis sibi nota sedilia nymphae.
Pastoresque latent, stertit sub sepe colonus,
Quis mihi blanditiasque tuas, quis tum mihi risus,   55
Cecropiosque sales referet, cultosque lepores?
Ite domum impasti, domino iam non vacat, agni.

At iam solus agros, iam pascua solus oberro,
Sicubi ramosae densantur vallibus umbrae,
Hic serum expecto, supra caput imber et Eurus   60
Triste sonant, fractaeque agitata crepuscula silvae.
Ite domum impasti, domino iam non vacat, agni.

Heu quam culta mihi prius arva procacibus herbis
Involvuntur, et ipsa situ seges alta fatiscit!
Innuba neglecto marcescit et uva racemo,   65
Nec myrteta iuvant; ovium quoque taedet, at illae
Moerent, inque suum convertunt ora magistrum.
Ite domum impasti, domino iam non vacat, agni.

Tityrus ad corylos vocat, Alphesiboeus ad ornos,
Ad salices Aegon, ad flumina pulcher Amyntas,   70
"Hic gelidi fontes, hic illita gramina musco,
Hic Zephyri, hic placidas interstrepit arbutus undas";
Ista canunt surdo, frutices ego nactus abibam.
Ite domum impasti, domino iam non vacat, agni.

49 desuper (from above) in contrast to heavenly love, at the poem's climax, who never shoots downwards. **52 umbra** (shade) either a literal shade, in which Pan hides, or Pan is 'hidden in', embodied in, the form of an oak. **55 blanditias** (coaxings) a word with sexual connotations. **56 Cecropios** for 'Attic salt', metaphor for elegant saltiness of wit, for which Athenians were famous. **59 ramosae umbrae** either branching shadows, or shadows of branches. **64 situ** (mould or rust)

the fierce wind throws all in confusion, and roars through
   the elms.
   Go homewards hungry, my lambs, your master is busy.   50

Or in summer at noon, the highest point of the day,
when secretive Pan takes naps in an oaken shade,
and the nymphs return to their places under the waters,
and shepherds take shelter, the farmer snores in a hedge,
can anyone bring you back, your beauty, your laughter,   55
your elegant biting wit, your culture, your charm?
   Go homewards hungry, my lambs, your master is busy.

But I wander alone through the fields now, alone through
   pastures;
wherever the branching shadows darken the valleys
I wait for the evening. The wind and the rain above me   60
moan, and twilight trembles in broken trees.
   Go homewards hungry, my lambs, your master is busy.

The fields that once I tended are wrapped in promiscuous
weeds, and the tall corn splinters with mould.
The vine is unweeded, and sags, its grapes unpicked,   65
the myrtles useless, delightless: I'm tired of my sheep
but they grieve, and turn their mouths towards their
   master.
   Go homewards hungry, my lambs, your master is busy.

Tityrus calls me to hazels, Alphesiboeus to ash trees,
Aegon to willows, and lovely Amyntas to streams: 70
"There are ice-cool fountains here, and moss-daubed
   banks,
and breezes, and arbutuses rustle by gentle streams."
They sing to the deaf. I reach some thickets and vanish.
   Go homewards hungry, my lambs, your master is busy.

the word suggests something that has been standing too long. **65
Innuba** (not wedded) to trees – the relevance to M's situation is obvious.
67 ora either faces, looking for guidance, or mouths, looking for food.
The refrain is given special relevance here.

Mopsus ad haec, nam me redeumtem forte notarat   75
(Et callebat avium linguas, et sydera Mopsus)
"Thyrsi quid hoc?" dixit, "quae te coquit improba bilis?
Aut te perdit amor, aut te male fascinat astrum,
Saturni grave saepe fuit pastoribus astrum,
Intimaque obliquo figit praecordia plumbo."   80
 Ite domum impasti, domino iam non vacat, agni.

Mirantur nymphae, et "quid te Thyrsi futurum est?
Quid tibi vis?" aiunt, "non haec solet esse iuventae
Nubila frons, oculique truces, vultusque severi,
Illa choros, lususque leves, et semper amorem   85
Iure petit, bis ille miser qui serus amavit."
 Ite domum impasti, domino iam non vacat, agni.

Venit Hyas, Dryopeque, et filia Baucidis Aegle
Docta modos, citharaeque sciens, sed perdita fastu,
Venit Idumanii Chloris vicina fluenti;   90
Nil me blanditiae, nil me solantia verba,
Nil me, si quid adest, movet, aut spes ulla futuri.
 Ite domum impasti, domino iam non vacat, agni.

Hei mihi quam similes ludunt per prata iuvenci,
Omnes unanimi secum sibi lege sodales,   95
Nec magis hunc alio quisquam secernit amicum
De grege, sic densi veniunt ad pabula thoes,
Inque vicem hirsuti paribus iunguntur onagri;
Lex eadem pelagi, deserto in littore Proteus
Agmina phocarum numerat, vilisque volucrum   100
Passer habet semper quicum sit, et omnia circum
Farra libens volitet, sero sua tecta revisens,
Quem si fors letho obiecit, seu milvus adunco

76 callebat basic meaning, to be callous, thick-skinned. Suggests Mopsa
is rendered insensitive by too much learning of a pragmatic kind.   **77
coquit** (cook, burn up) technical term from the medical theory of
humours.   **78 Aut** (either) he immediately answers his own question.
80 plumbo (lead) the metal of Saturn in alchemical theory.   **86 iure** (by
law) the law of youth; cf *El VII* 1.   **90 Idumanii** refers to Blackwater
river and estuary, in Essex. Hence presumably a specific girl.   **94 iuvenci**

Then Mopsus spoke, for he happened to see me return, 75
that expert decipherer of stars and the language of birds:
"Thyrsis, what's this? What evil humour torments you?
It's love that destroys you, or spells from a hostile star.
The planet Saturn is often adverse to shepherds,
his leaden shaft can angle deep in their breast." 80
  Go homewards hungry, my lambs, your master is busy.

The nymphs are puzzled: "Thyrsis, what will become of
      you?
What do you want? It's abnormal for youth to be thus:
gloomy of brow, eyes grim, and face so stern.
Youth likes dancing and pleasant sports, and its law 85
is always to love. And love delayed has twice the pain."
  Go homewards hungry, my lambs, your master is busy.

And Hÿäs and Drÿopë and Baucis' daughter Aeglé
(a skilful musician and poet, but ruined by pride)
and Chloris from nearby Blackwater Bay, all came, 90
but their flirting and comforting words meant nothing,
      nothing,
nothing can move me, the present or hopes for the future.
  Go homewards hungry, my lambs, your master is busy.

Young bulls that play in the fields are all alike,
companions all, with a single mind and law, 95
and none selects just one particular friend
from the herd: and wolves attack their food in packs,
and the long-haired mountain asses are joined in pairs.
The law of the sea is the same, where Proteus orders
his seals on a lonely beach in troops. And the sparrow, 100
that humblest of birds, is always accompanied, and flies
round crops as it likes, and returns at night to its nest.
But if chance brings death – a kite with its curved beak

young bulls – though the male gender might include heifers. Throughout,
M underplays heterosexual mating in nature. **99 Proteus** god of the sea,
Neptune's herdsman, famous for being able to change his shape at will,
this quality perhaps relevant to the changeability of animals' affections.

Fata tulit rostro, seu stravit arundine fossor,
Protinus ille alium socio petit inde volatu.　　　　　105
Nos durum genus, et diris exercita fatis
Gens homines aliena animis, et pectore discors,
Vix sibi quisque parem de millibus invenit unum,
Aut si sors dederit tandem non aspera votis,
Illus inopina dies qua non speraveris hora　　　　110
Surripit, aeternum linquens in saecula damnum.
　Ite domum impasti, domino iam non vacat, agni.

Heu quis me ignotas traxit vagus error in oras
Ire per aereas rupes, Alpemque nivosam!
Ecquid erat tanti Romam vidisse sepultam?　　　　115
Quamvis illa foret, qualem dum viseret olim,
Tityrus ipse suas et oves et rura reliquit;
Ut te tam dulci possem caruisse sodale,
Possem tot maria alta, tot interponere montes,
Tot sylvas, tot saxa tibi, fluviosque sonantes.　　120
Ah certe extremum licuisset tangere dextram,
Et bene compositos placide morientis ocellos,
Et dixisse "vale, nostri memor ibis ad astra."
　Ite domum impasti, domino iam non vacat, agni.

Quamquam etiam vestri nunquam meminisse pigebit　125
Pastores Thusci, musis operata iuventus,
Hic charis, atque lepos; et Thuscus tu quoque Damon.
Antiqua genus unde petis Lucumonis ab urbe.
O ego quantus eram, gelidi cum stratus ad Arni
Murmura, populeumque nemus, qua mollior herba,　130
Carpere nunc violas, nunc summas carpere myrtos,
Et potui Lycidae certantem audire Menalcam.
Ipse etiam tentare ausus sum, nec puto multum

105 alium (another) masculine gender again. **113 error** means both
a wandering, and an error, so that to wander *is* an error. This
double sense reinforced by **vagus**, also meaning wandering. **115
sepultam** (buried) Rome's remains like a corpse, like Damon. **122
morientis** (dying) present participle, suggesting Damon is still just
alive as M performs this rite. **123 nostri** (us) either 'our friendship',

the agent of fate, or a labourer who shoots it down –
the other will find a companion to fly with soon.   105
We men are callous, a species vexed by a curse.
Minds are estranged from minds, and hearts from hearts.
It's hard to discover a single friend from thousands;
but if fate should yield to your prayers, and grant such a
    one,
a day and an hour not expected or hoped for takes him,   110
leaving a lifetime of pain, an eternal loss.
  Go homewards hungry, my lambs, your master is busy.

What straying fancies induced me to wander abroad,
and cross the windswept crags of the snowy alps?
Did I want so much to gaze at the relics of Rome,   115
even a Rome as it was in days of old
when Tityrus left his flocks and fields to see it,
that I brought myself to leave you, so sweet, a friend,
and set such seas between us, so many mountains,
so many forests and crags and echoing streams?   120
I'm sure they'd have let me touch his hand at the end,
and carefully close his eyes as he died in peace,
and say: "Farewell! remember your friend as you go to
    the stars."
  Go homewards hungry, my lambs, your master is busy.

But still, I'll never regret my memories of you,   125
you Tuscan shepherds, youths who worship the muse,
and your grace and charm (you too were a Tuscan, Damon,
and traced your line from the ancient city of Lucca).
How exalted I felt, as I stretched by the ice-cool Arno's
murmurs, on soft grass in a poplar grove,   130
and plucked a violet, or plucked a myrtle sprig,
and heard Manálcas and Lycidas compete.
And I dared to try my skill, and I think I didn't

or plural with singular meaning, 'me'. **126 iuventus** (youths) conventional. Manso was nearly 70. **128 Lucumonis** Lucca was in
Tuscany. **129 eram** (I was) not 'I felt important' but 'I was'. **132**
reference to poetry readings in Italian academies.

Displicui, nam sunt et apud me munera vestra
Fiscellae; calathique et cerea vincia cicutae,   135
Quin et nostra suas docuerunt nomina fagos
Et Datis, et Francinus, erant et vocibus ambo
Et studiis noti, Lydorum sanguinis ambo.
  Ite domum impasti, domino iam non vacat, agni.

Haec mihi tum laeto dictabat roscida luna,   140
Dum solus teneros claudebam cratibus hoedos.
Ah quoties dixi, cum te cinis ater habebat,
"Nunc canit, aut lepori nunc tendit retia Damon,
Vimina nunc texit, varios sibi quod sit in usus";
Et quae tum facili sperabam mente futura   145
Arripui voto levis, et praesentia finxi,
"Heus bone numquid agis? nisi te quid forte retardat,
Imus? et arguta paulum recubamus in umbra,
Aut ad aquas Colni, aut ubi iugera Cassivelauni?
Tu mihi percurres medicos, tua gramina, succos,   150
Helleborumque, humilesque crocos, foliumque
      hyacinthi,
Quasque habet ista palus herbas, artesque medentum",
Ah pereant herbae, pereant artesque medentum
Gramina, postquam ipsi nil profecere magistro.
Ipse etiam, nam nescio quid mihi grande sonabat   155
Fistula, ab undecima iam lux est altera nocte,
Et tum forte novis admoram labra cicutis,
Dissiluere tamen rupta compage, nec ultra
Ferre graves potuere sonos, dubito quoque ne sim
Turgidulus, tamen et referam, vos cedite silvae.   160
  Ite domum impasti, domino iam non vacat, agni.

135 Fiscellae shepherd's baskets etc stand for poetic offerings.   **137** Dati
and Francini both wrote complimentary verses for M, which he reprinted
as prefaces to his Latin verses.   **144 usus** (purposes, uses) thinking ahead
on Damon's behalf, to what he will use these baskets for.   **147 retardat**
(hinders) ominous.   **149** Cassivelaunus' domain included Horton, M's
family home. There is a river Colne, tributary of the Thames, nearby.
154 magistro (master) both were masters of their arts. Damon's medical
arts have failed him, and M's poetic powers seem to be breaking under
the strain. But by a free association process through **ipsi** (self) and

displease you, for here beside me I have your gifts,
baskets and bowls, and pipes bonded with wax.   135
Francini and Dati inscribed my name on beeches,
both of them famous poets
and scholars, and both of Etruscan blood.
    Go homewards hungry, my lambs, your master is busy.

The dew-bright moon would tell me these triumphs, as 140
    happy,
alone, I'd shut my delicate kids in their pens,
How often I said – when the black grave already possessed
    you –
"Damon is singing now, or laying traps for a hare,
and now he's weaving baskets for various uses."
Whatever I hoped in my facile mind would happen,   145
I lightly seized on the wish, and imagined it true.
"Ah friend! what's doing? If nothing stands in the way,
let's go and rest for a while in a sharp-etched shade
by the river Colne, or in Cassivellaunus' domain.
You'll tell me your stock of healing potions and plants,   150
hellebore, lowly crocuses, hyacinth leaves,
the herbs that grow in the marshes, your medical
    skills."
O death to all herbs! And death to medical skills,
and plants, for they failed to cure the master himself!
Myself I! – for I don't know what grand theme my   155
    pipes
were sounding – eleven nights to the light of this day –
then maybe the pipes were new that I put to my lips,
but they shattered, broke at the joins, no longer able
to take so powerful a sound; I pause lest I seem
puffed up, but I'll sing it again: so farewell, you woods. 160
    Go homewards hungry, my lambs, your master is busy.

master he gets to himself and his own mastery. **156 lux** (light) means
both this day, and the light, the break-through after his long nights of
grief. **160 cedite** (yield, give way) formula for comparison between
poetic genres (see eg *El I* 23) signals M's rejection of the pastoral genre.

Ipse ego Dardanias Rutupina per aequora puppes
Dicam, et Pandrasidos regnum vetus Inogeniae,
Brennumque Arviragumque duces, priscumque Belinum,
Et tandem Armoricos Britonum sub lege colonos;   165
Tum gravidam Arturo fatali fraude Iogernen
Mendaces vultus, assumptaque Gorlois arma,
Merlini dolus. O mihi tum si vita supersit,
Tu procul annosa pendebis fistula pinu
Multum oblita mihi, aut patriis mutata camoenis   170
Brittonicum strides, quid enim? omnia non licet uni
Non sperasse uni licet omnia, mi satis ampla
Merces, et mihi grande decus (sim ignotus in aevum
Tum licet, externo penitusque inglorius orbi)
Si me flava comas legat Usa, et potor Alauni,   175
Vorticibusque frequens Abra, et nemus omne Treantae,
Et Thamesis meus ante omnes, et fusca metallis
Tamara, et extremis me discant Orcades undis.
   Ite domum impasti, domino iam non vacat, agni.

Haec tibi servabam lenta sub cortice lauri,   180
Haec, et plura simul, tum quae mihi pocula Mansus,
Mansus Chalcidicae non ultima gloria ripae
Bina dedit, mirum artis opus, mirandus et ipse,
Et circum gemino caelaverat argumento:
In medio rubri maris unda, et odoriferum ver   185
Littora longa Arabum, et sudantes balsama silvae,
Has inter Phoenix divina avis, unica terris
Caeruleum fulgens diversicoloribus alis
Auroram vitreis surgentem respicit undis.
Parte alia polus omnipatens, et magnus Olympus,   190

162 some allusions to early British history, mainly from Geoffrey of
Monmouth's very unreliable history. **166** this episode might have
some relevance to M's situation. The widowed Igraine (like the widowed
M?) was tricked into a most auspicious copulation. **169 procul annosa**
(far off, ancient) Latin language and poetic forms, which M will renounce
unless they can be adapted to British subject matter. **172 sperasse**
(hope) self-control includes limiting one's hopes as well as desires. **181**
pocula (goblets) stand for books by Manso, perhaps his neoplatonic
dialogues about love and beauty, and his translation of Claudian's

I'll sing of the Trojan ships off the coast of Kent,
and the ancient kingdom of Inogen, Pandrasus' daughter,
lords Brennus and Arviragus, and ancient Belinus,
then those who settled in France under British rule: 165
then Igraine, pregnant with Arthur, deceived by Uther
disguised with the face of Gorlois, and clothed in his
        armour,
by Merlin's magical arts. If I live so long,
you'll hang from some far-off ancient pine, my pipes,
dim in my memory, unless, transformed by my native   170
        Muse,
you trumpet a British theme. Why not? No man
is allowed to do all, or hope to. I'll count it sufficient
reward and ample honour if – unknown for ever
perhaps, not famous at all in the outside world –
the yellow-haired Ouse should read me, and Avonside   175
        dwellers,
and eddying Humber, and all the woods by the Trent,
and especially my Thames, and the Tamar, dark with its
        ores,
and the distant waves of the Orkneys, all learn my songs.
    Go homewards hungry, my lambs, your master is busy.

I was keeping you these, preserved in dense laurel-bark, 180
and similar treasures: the goblets which Manso gave me,
yes, Manso, not least of the glories of Naples' shores;
two goblets, the wonderful works of a wonderful man.
Each was embossed around with a double theme.
In the middle the Red Sea, and spring with her scents,   185
the long Arabian coast, and trees dripping with balm;
with a phoenix, heavenly bird, just one in the world,
resplendently blue, with wings of myriad hues,
watching the dawn ascend from a glassy sea.
And opposite, heaven was opening, and great Olympus,   190

Phoenix. **185** the Red Sea stands for the boundary between life and
death, beyond which is paradise for Damon. The phoenix, which is
reborn from its own ashes, is a symbol of the resurrection. **189**
surgentem (rising) another resurrection image.

Quis putet? hic quoque Amor, pictaeque in nube pharetrae,
Arma corusca faces, et spicula tincta pyropo;
Nec tenues animas, pectusque ignobile vulgi
Hinc ferit, at circum flammantia lumina torquens
Semper in erectum spargit sua tela per orbes   195
Impiger, et pronos nunquam collimat ad ictus,
Hinc mentes ardere sacrae, formaeque deorum.

   Tu quoque in his, nec me fallit spes lubrica Damon,
Tu quoque in his certe es, nam quo tua dulcis abiret
Sanctaque simplicitas, nam quo tua candida virtus?   200
Nec te Lethaeo fas quaesivisse sub Orco,
Nec tibi conveniunt lacrymae, nec flebimus ultra,
Ite procul lacrymae, purum colit aethera Damon,
Aethera purus habet, pluvium pede reppulit arcum;
Heroumque animas inter, divosque perennes,   205
Aethereos haurit latices et gaudia potat
Ore sacro. Quin tu coeli post iura recepta
Dexter ades, placidusque fave quicunque vocaris,
Seu tu noster eris Damon, sive aequior audis
Diodotus, quo te divino nomine cuncti   210
Coelicolae norint, sylvisque vocabere Damon.

Quod tibi purpureus pudor, et sine labe iuventus
Grata fuit, quod nulla tori libata voluptas,
En etiam tibi virginei servantur honores;
Ipse caput nitidum cinctus rutilante corona,   215
Letaque frondentis gestans umbracula palmae
Aeternum perages immortales hymenaeos;
Cantus ubi, choreisque furit lyra mista beatis,
Festa Sionaeo bacchantur et orgia thyrso.'

191 **Amor** (love) has all the traditional trappings of Cupid or Eros. **201**
fas (right, permissible). It's not only painful but wrong to look for Damon
below. **203 purum** (pure) play on this word, describing first the sky then
Damon, has the effect of assimilating him to the sky. **206 haurit** (quaffs)
Dionysiac response to joys. **207 iura recepta** (the prescribed rites
being fulfilled). This releases Damon to become a guardian spirit to his
friend. **210 Diodotus** means god given, so this is a pun on his name.
212–14 see *Rev* xiv 'These are they which were not defiled with women;

and, amazing! Love in a cloud, with painted quiver,
glittering armour, torch and bronze-tipped darts.
Not striking the feeble of soul, or the squalid breasts
of the masses, but rolling his flaming eyes around,
he scatters his arrows upwards towards the spheres   195
unwearied, and never directs his missiles to earth,
so holy minds are inflamed, and appear like gods.

You too, my Damon – no specious hopes deceive me –
you surely were there in that host. For where would your
     sweet
and holy simplicity go, your radiant virtue?   200
It's wrong to seek you below, in the realms of the dead,
and tears are unfitting, nor will I weep any more.
So away with tears, for Damon dwells in the pure
heavens, pure as the heavens, and treads on the rainbow.
He stands with the souls of heroes, immortal gods,   205
draining the liquors of heaven, and drinking delight
with holy lips. Now heaven has formally claimed you,
come close, and gently sustain me; whatever your name
     now,
the Damon we called you, or else more properly titled
Diodati, the god-given name which the hosts of heaven   210
all know, in the woods your name will still be Damon.

You delighted in modesty's blush, and a youth without
     stain,
and never tasted the joys of the marriage bed,
so look! the honours due to a virgin await you!
Your radiant head is crowned with a glittering garland,   215
you carry luxuriant branches of living palm,
and enact for ever the rites of the deathless marriage,
and songs and music madden the dancing saints,
and the furious orgies rage to the thyrsus of Zion.'

for they are virgins.' **216-17** more images from *Revelation*. **217**
hymenaeos the marriage of the Lamb (Christ) *Rev* xix. There the
marriage is followed immediately by war and destruction. In *Damon*
the marriage celebrations go on for ever (*aeternum*).

Sonnet 7 *How soon hath time*

HOW SOON HATH TIME the subtle thief of youth,
    Stol'n on his wing my three and twentieth year!
    My hasting days fly on with full career,
But my late spring no bud or blossom sheweth.
Perhaps my semblance might deceive the truth,   5
    That I to manhood am arrived so near,
    And inward ripeness doth much less appear,
    That some more timely-happy spirits endueth.
Yet be it less or more, or soon or slow,
    It shall be still in strictest measure even,   10
    To that same lot, however mean or high,
Toward which time leads me, and the will of heaven;
    All is, if I have grace to use it so,
    As ever in my great task-master's eye.

Sonnet 8 *When the assault was intended to the city*

CAPTAIN OR COLONEL, or knight in arms,
    Whose chance on these defenceless doors may
        seize,
If deed of honour did thee ever please,
Guard them, and him within protect from harms,
He can requite thee, for he knows the charms   5
    That call fame on such gentle acts as these,
    And he can spread thy name o'er lands and seas,
    Whatever clime the sun's bright circle warms.

[*Sonnet 7*] **5 semblance** appearance. M looked young for his age. **deceive the truth** deceive the eyes and conceal the truth, or, trick the truth and give me more time. **6 near** at 23–24 he would have been very near. Does this betray his anxieties? **7 appear** show on the outside. **8 timely-happy** happy or fortunate (Lat *felix*) at the proper time. **endueth** clothes. Emphasizing how external their 'maturity' is. But cf *El V* 3. **9–14** see Commentary p. 40. **10 still** always. [*Sonnet 8*] **5 charms** Latin *carmen* meant both spell or incantation, and poem. **6 gentle** both noble and mild, a significant double meaning suggesting that to be noble is to be mild. **8 sun** Apollo as god of the sun and of poetry can ensure this.

Lift not thy spear against the muses' bower,
   The great Emathian conqueror bid spare   10
   The house of Pindarus, when temple and tower
Went to the ground: and the repeated air
   Of said Electra's poet had the power
   To save the Athenian walls from ruin bare.

Sonnet 9   *Lady, that in the prime*

LADY, THAT IN THE PRIME of earliest youth,
   Wisely hath shunned the broad way and the green,
   And with those few art eminently seen,
That labour up the hill of heavenly truth,
The better part with Mary and with Ruth,   5
   Chosen thou hast, and they that overween,
   And at thy growing virtues fret their spleen,
   No anger find in thee, but pity and ruth.
Thy care is fixed and zealously attends
   To fill thy odorous lamp with deeds of light,   10
   And hope that reaps not shame. Therefore be sure
Thou, when the bridegroom with his feastful friends
   Passes to bliss at the mid-hour of night,
   Hast gained thy entrance, virgin wise and pure.

Sonnet 10   *To the Lady Margaret Ley*

DAUGHTER TO THAT GOOD EARL, once president
   Of England's council, and her treasury,
   Who lived in both, unstained with gold or fee.
And left them both, more in himself content,

10 **Emathian** Alexander the Great. 11 Pindar, famous poet of early
Greece. 12–14 after the Spartans had defeated Athens in 404 BC, the
city was spared after someone was heard singing a chorus from Euripides'
Electra. Euripides was recently dead, an opponent of the war. [*Sonnet*
9] 2 **broad way** that leads to destruction (*Matt* vii). **green** not clear
what text or significance M had in mind. Is M against grass, and things
that grow? 8 **ruth** pity – M repeating himself for the pun/rhyme?
9 **attends** waits. 11–14 parable of the ten virgins, *Matt* xxv.

Till the sad breaking of that parliament   5
    Broke him, as that dishonest victory
    At Chaeronea, fatal to liberty
    Killed with report that old man eloquent,
Though later born, than to have known the days
    Wherein your father flourished, yet by you,   10
    Madam, methinks I see him living yet;
So well your words his noble virtues praise,
    That all both judge you to relate them true,
    And to possess them, honoured Margaret.

Sonnet 11   *A book was writ of late*

A BOOK WAS WRIT OF LATE called *Tetrachordon*;
    And woven close, both matter, form and style;
    The subject new: it walked the town awhile,
Numbering good intellects; now seldom pored on.
Cries the stall-reader, Bless us! what a word on   5
    A title-page is this! And some in file
    Stand spelling false, while one might walk to Mile-
End Green. Why is it harder sirs than Gordon,
Colkitto, or Macdonnel, or Galasp?
    Those rugged names to our like mouths grow sleek   10
    That would have made Quintilian stare and gasp.
Thy age, like ours, O soul of Sir John Cheke,
    Hated not learning worse than toad or asp;
    When thou taught'st Cambridge, and King
        Edward Greek.

[*Sonnet 10*] **5** Charles forcibly dissolved parliament 2 Mar 1629. Ley died 14 Mar aged 79. **7 Chaeronea** where the tyrant Philip of Macedon, father of Alexander the Great, defeated the Athenians. The orator Isocrates, then aged 98, died two days later, having vowed that he would not live in an unfree Athens. [*Sonnet 11*] **1 Tetrachordon** title of M's third tract on divorce, published Mar 1645. The Greek words mean 'four chords/strings', hence perhaps *woven close*. **7 spelling false** misinterpreting. **8** names of Scottish presbyterians, probably chosen for sound rather than political importance. **10 like** similarly rugged. **11 Quintilian** Roman authority on rhetoric, criticized the importing of foreign words. **12 like** M may be getting his negatives confused, if

Sonnet 12   On the detraction which followed
upon my writing certain treatises

I DID BUT PROMPT THE AGE to quit their clogs
 By the known rules of ancient liberty,
 When straight a barbarous noise environs me
Of owls and cuckoos, asses, apes and dogs.
As when those hinds that were transformed to frogs   5
 Railed at Latona's twin-born progeny
 Which after held the sun and moon in fee.
But this is got by casting pearl to hogs;
 That bawl for freedom in their senseless mood,
 And still revolt when truth would set them free.   10
 Licence they mean when they cry liberty;
For who loves that, must first be wise and good;
 But from that mark how far they rove we see
 For all this waste of wealth, and loss of blood.

On the new forcers of conscience
under the Long Parliament

BECAUSE YOU HAVE THROWN OFF your prelate
 lord,
 And with stiff vows renounced his liturgy
To seize the widowed whore plurality
From them whose sin ye envied, not abhorred,
Dare ye for this adjure the civil sword   5

he means that the age of Cheke (first professor of Greek at Cambridge,
1540–51) was *un*like ours. This was the early period of English humanism
and protestantism, which M may be looking back to as a golden age.
But in fact there was opposition to both Greek and the Reformation
even in Cheke's day, so M might really mean 'like'. [*Sonnet 12*] **5**
Apollo and his sister Diana. While they were still infants, their mother
Latona was fleeing with them from jealous Juno. Some peasants prevented
them from drinking at a spring, and were turned into frogs by Jove
(Ovid *Met* VI 317). [*New forcers*] **1 prelate** bishop. **lord** perhaps a
pun on the name of Archbishop Laud. **3 plurality** the custom of
receiving income from more than one parish.

To force our consciences that Christ set free,
And ride us with a classic hierarchy
Taught ye by mere A. S. and Rutherford?
Men whose life, learning, faith and pure intent
  Would have been held in high esteem with Paul   10
Must now be named and printed heretics
By shallow Edwards and Scotch What-d'ye-call:
  But we do hope to find out all your tricks,
  Your plots and packing worse than those of Trent,
     That so the parliament   15
May with their wholesome and preventive shears
Clip your phylacteries, though baulk your ears,
     And succour our just fears
When they shall read this clearly in your charge
New *Presbyter* is but old *Priest* writ large.   20

Sonnet 13   To Mr H. Lawes, on his airs

HARRY, WHOSE TUNEFUL and well-measured song
    First taught our English music how to span
    Words with just note and accent, not to scan
With Midas' ears, committing short and long;

6 our includes only M and his fellow saints, since Christ set free only
the consciences of the elect, not everyone's. **7 classic** the unit of organization in the presbyterian church was called the 'classis'. **hierarchy** means
rule by priests. **8** Adam Stewart and Rutherford were two leading presbyterians. **12** Thomas Edwards, vehement writer of pamphlets, his
Gangraena of 1646 attacked M among others. **14 packing** organizing
majorities. **Trent** the catholic council of Trent (1545–63), which worked
out Counter-reformation doctrine and policy. M consistently associating
the so-called 'reformed' church with their hated enemy papism, through
their common political motives and tactics. **17 phylacteries** small
leather container of sacred texts from OT, worn round neck by pious
Jews. Used as symbol of pharisees' hypocrisy by Christ *Matt* xxiii.
baulk a veiled reference to the puritan extremist, William Prynne, whose
ears were mutilated by order of Charles, because of alleged libels
against the king and queen. M originally wrote 'crop ye as close as
marginal P-----'s ears', which would have seemed to endorse Charles'
brutal act of censorship. **19 charge** bill of accusation. [*Sonnet 13*]
2 First flattery, since the great Elizabethan composers had done so
earlier. **4** Midas judged Pan's music was better than Apollo's, who
gave him ass's ears.

Thy worth and skill exempts thee from the throng,　　5
　　With praise enough for envy to look wan;
　　To after age thou shalt be writ the man,
　　That with smooth air couldst humour best our tongue.
Thou honour'st verse, and verse must lend her wing
　　To honour thee, the priest of Phoebus' choir　　10
　　That tun'st their happiest lines in hymn, or story.
Dante shall give fame leave to set thee higher
　　Than his Casella, whom he wooed to sing
　　Met in the milder shades of Purgatory.

Sonnet 14　When faith and love

WHEN FAITH AND LOVE which parted from thee
　　never,
　　　　Had ripened thy just soul to dwell with God,
Meekly thou didst resign this earthy load
Of death, called life; which us from life doth sever.
Thy works and alms and all thy good endeavour　　5
　　Stayed not behind, nor in the grave were trod;
　　But as faith pointed with her golden rod,
　　Followed thee up to joy and bliss for ever.
Love led them on, and faith who knew them best
　　Thy handmaids, clad them o'er with purple beams　　10
　　And azure wings, that up they flew so dressed,
And speak the truth of thee on glorious themes
　　Before the judge, who thenceforth bid thee rest
　　And drink thy fill of pure immortal streams.

On the Lord General Fairfax at the siege of Colchester

FAIRFAX, WHOSE NAME in arms through Europe
　　rings
　　　　Filling each mouth with envy, or with praise,

12–14 Dante met the Florentine composer in Purgatory (*Purg* XI 76–119).
[*Sonnet 14*] **7** contrast with the golden rod of *Damon* 23, pagan
and a rod of justice. **14** weak echo of *Damon* 206.

And all her jealous monarchs with amaze,
And rumours loud, that daunt remotest kings,
Thy firm unshaken virtue ever brings   5
    Victory home, though new rebellions raise
    Their hydra heads, and the false North displays
    Her broken league, to imp their serpent wings,
O yet a nobler task awaits thy hand;
    For what can war, but endless war still breed,   10
    Till truth, and right from violence be freed,
And public faith cleared from the shameful brand
Of public fraud. In vain doth valour bleed
While avarice, and rapine share the land.

To the Lord General Cromwell

CROMWELL, OUR CHIEF OF MEN, who through a
        cloud
        Not of war only, but detractions rude,
Guided by faith and matchless fortitude
To peace and truth thy glorious way has ploughed,
And on the neck of crowned fortune proud   5
    Hast reared God's trophies and his work pursued,
    While Darwen stream with blood of Scots imbrued,
    And Dunbar field resounds thy praises loud,
And Worcester's laureate wreath; yet much remains
    To conquer still; peace hath her victories   10

[*On Fairfax*] **7 hydra** monster killed by Hercules, its heads grew
immediately they were cut off. A common symbol of rebellion. **false
North** Scotland who though presbyterian aided a royalist invasion
of England in 1648. **8 imp** graft new feathers onto stumps of old ones.
M's metaphors out of control, since the hydra had no wings, and serpents
have no feathers. [*To Cromwell*] **2 detractions** the problems of peace
(paired terms based on peace–war thread through the poem; or more
exactly, religious and physical orientation, success in things spiritual to
match Cromwell's secular achievements). **4 ploughed** but no crop is
envisaged. **5–6** metaphors mixed almost ludicrously – especially the
image of Cromwell 'pursuing his work' round the neck of Fortune.
7 Darwen battle of Preston, 17 Aug 1648, where Cromwell defeated the
Scots. **8 Dunbar** another victory against the Scots, 1650. **9 Worcester**
1651. All these were victories against the Scots, external enemies, and
renegade presbyterians.

No less renowned than war, new foes arise
Threatening to bind our souls with secular chains:
Help us to save free conscience from the paw
Of hireling wolves whose gospel is their maw.

To Sir Henry Vane the younger

VANE, YOUNG IN YEARS, but in sage counsel old,
Than whom a better senator ne'er held
The helm of Rome, when gowns not arms repelled
The fierce Epirot and the African bold.
Whether to settle peace or to unfold   5
The drift of hollow states, hard to be spelled,
Then to advise how war may best, upheld,
Move by her two main nerves, iron and gold
In all her equipage: besides to know
Both spiritual power and civil, what each means,   10
What severs each, thou hast learned, which few have
      done.
The bounds of either sword to thee we owe;
Therefore on thy firm hand Religion leans
In peace, and reckons thee her eldest son.

Sonnet 15   On the late massacre in Piedmont

AVENGE O LORD thy slaughtered saints, whose
      bones
Lie scattered on the Alpine mountains cold,

14 **hireling wolves** compressed construction, for hireling shepherds
who are really wolves. [*To Vane*] 4 **Epirot** Pyrrhus invaded Italy
281 BC. He won a battle but suffered such heavy casualities that it
was worse than a defeat. **African** Hannibal, ruthless Carthaginian
general, whose invasion of Italy in 218 was militarily brilliant but
failed for lack of political organization. 6 **hollow** pun on Holland,
with whom Vane had just been negotiating. The Anglo-Dutch war
broke out shortly before the composition of this sonnet. 8 **nerves**
sinews. **iron** military equipment. 11 **severs each** shows in what way
they are different: but the metaphor suggests the action of yet another
sword, which will cut between the other two. 13 **hand** for arm. 14
In peace only in peace, not in war? or peacefully?

Even them who kept thy truth so pure of old
When all our fathers worshipped stocks and stones,
Forget not: in thy book record their groans   5
    Who were thy sheep and in their ancient fold
    Slain by the bloody Piedmontese that rolled
    Mother with infant down the rocks. Their moans
The vales redoubled to the hills, and they
    To heaven. Their martyred blood and ashes sow   10
    O'er all the Italian fields where still doth sway
The triple tyrant: that from these may grow
    A hundredfold, who having learnt thy way
    Early may fly the Babylonian woe.

Sonnet 16   *When I consider*

WHEN I CONSIDER how my light is spent,
    Ere half my days, in this dark world and wide,
    And that one talent which is death to hide,
    Lodged with me useless, though my soul more bent
To serve therewith my maker, and present   5
    My true account, lest he returning chide,
    Doth God exact day-labour, light denied,
    I fondly ask; but Patience to prevent
That murmur, soon replies, God doth not need
    Either man's work or his own gifts, who best   10
    Bear his mild yoke, they serve him best, his state
Is kingly. Thousands at his bidding speed
    And post o'er land and ocean without rest:
    They also serve who only stand and wait.

[*Sonnet 15*]   **12 triple tyrant** the papacy, identified with corrupt Babylon
of *Rev* xiv, xvii, xviii.   [*Sonnet 16*]   **11 yoke** cf *Matt* xi 30 'my yoke
is easy'.

Sonnet 17   *Lawrence of virtuous father*

L AWRENCE OF VIRTUOUS FATHER virtuous son,
　　Now that the fields are dank, and ways are mire,
　　Where shall we sometimes meet, and by the fire
Help waste a sullen day; what may be won
From the hard season gaining: time will run   5
　　On smoother, till Favonius reinspire
　　The frozen earth; and clothe in fresh attire
The lily and rose, that neither sowed nor spun.
What neat repast shall feast us, light and choice,
　　Of Attic taste, with wine, whence we may rise   10
　　To hear the lute well touched, or artful voice
Warble immortal notes and Tuscan air?
　　He who of those delights can judge, and spare
　　To interpose them oft, is not unwise.

Sonnet 18   *Cyriack, whose grandsire*

C YRIACK, WHOSE GRANDSIRE on the royal bench
　　Of British Themis, with no mean applause
　　Pronounced and in his volumes taught our laws,
Which others at their bar so often wrench;
Today deep thoughts resolve with me to drench   5
　　In mirth, that after no repenting draws;
　　Let Euclid rest and Archimedes pause,

[*Sonnet 17*] **2 fields** M was probably living in London, so this is a
mutedly pastoral image (cf *Damon*). **6 Favonius** west wind. **8 lily**
allusion to Christ's words, 'Why take ye thought for raiment? Consider
the lilies of the field, how they grow; they toil not, neither do they
spin' (*Matt* vi). Christ didn't mention roses, conventional symbols of
passion. **10 Attic** the cultivated taste of the Greeks. Recalls reference
to Diodati in *Damon* 56. **13 spare** syntax is obscure: either 'spare
time to' ie indulge frequently, or 'be sparing of' ie judgment, con-
noisseurship is dominant. If the second, M has changed direction
sharply, but if the first, why does he put it so obscurely? [*Sonnet 18*]
1 grandsire Sir Edward Coke, famous lawyer, who opposed James' and
Charles' assertions of the royal prerogative. **7 Euclid** geometry.
Archimedes physics.

And what the Swede intend, and what the French.
To measure life, learn thou betimes, and know
Toward solid good what leads the nearest way;   10
For other things mild heaven a time ordains,
And disapproves that care, though wise in show,
That with superfluous burden loads the day,
And when God sends a cheerful hour, refrains.

To Mr Cyriack Skinner upon his blindness

CYRIACK, THIS THREE YEARS' DAY these eyes,
    though clear
    To outward view, of blemish or of spot;
Bereft of light their seeing have forgot,
Nor to their idle orbs doth sight appear
Of sun or moon or star throughout the year,   5
Or man or woman. Yet I argue not
Against heaven's hand or will, nor bate a jot
Of heart or hope; but still bear up and steer
Right onward. What supports me dost thou ask?
The conscience, friend, to have lost them overplied   10
In liberty's defence, my noble task,
Of which all Europe talks from side to side.
This thought might lead me through the world's vain mask
Content though blind, had I no better guide.

Sonnet 19   Methought I saw

METHOUGHT I SAW my late espoused saint
    Brought to me like Alcestis from the grave,
    Whom Jove's great son to her glad husband gave,

9–10 maths applied to ethics (as Hobbes was notorious for doing, eg in
his *Leviathan* of 1651). [*To Mr Cyriack Skinner*] 1 day anniversary of
the day he went blind; also suggesting that this whole period has been a
continuous day, not a continuous night as others might think. 10 conscience consciousness. He lost his sight finally while working, against
doctor's orders, on his *First defence* of 1651. 13 mask courtly play in
which the performers were masked. [*Sonnet 19*] 1 late recently married:
to him (Katherine was married to him for only 15 months before she died)
but also married to Christ.

Rescued from death by force though pale and faint.
Mine as whom washed from spot of childbed taint,   5
   Purification in the old laws did save,
   And such, as yet once more I trust to have
Full sight of her in heaven without restraint,
Came vested all in white, pure as her mind:
   Her face was veiled, yet to my fancied sight,   10
   Love, sweetness, goodness in her person shined
So clear, as in no face with more delight.
   But O as to embrace me she inclined
   I waked, she fled, and day brought back my night.

6 laws of OT, now superseded under Christ's dispensation.   **7 once more** he had never seen Katherine.   **10 fancied sight** the sight I imagined I had.

Introduction to Samson Agonistes

Reading

As usual in this series, spelling is modernized but punctuation original, except in a few cases where it would distort the sense. But don't take the punctuation as a set of rules: it is unreliable; some of its conventions are alien to us (Milton's colons and semi-colons, semi-colons and commas, are mixed by our usage); and in any case its virtue is its laxity. It allows you – requires you – to hear the intonation and the rhythms from the meaning, and so put in your own punctuation. When the Chorus tell Samson that the next visitor is 'Dálila thy wife' (724) he says, 'My wife, my traitress, let her not come near me'. Obviously that has to be interpreted as 'My wife? my *traitress*! Let her not come near me!' Listen for emphasis. It often comes in lists of similar words, eg Delilah's unpunctuated pleading promise, that ends in a little stab of admitted guilt and assumed excuse:

> *Let* me obtain forgiveness of thee, Samson...
> Afford me place to show what recompense
> Towards thee I intend for what I have *misdone*,
> *Misguided*;   909

Notice the variation of stress in some of these lists, eg 'ig*no*ble, *Un*manly, igno*min*ious, *in*famous' (416). Lists often give the verse a rhetorical patterning that holds it up against its own freedom, eg

> Weakness is thy excuse,
> And I believe it, weakness to resist
> Philístian gold: if weakness may excuse,
> What murderer, what traitor, parricide,
> Incestuous, sacrilegious, but may plead it?   829

The question there is another feature of the poem; and you have to ask what the listings, the rhetoric, and the queryings represent as parts of the total meaning.

The verse is rhetorical in the sense of using 'schemes', syntactical patterns, repetitions, and rhymes, rather than metaphors (see section on rhetoric in *PL: introduction* in this series). Eg:

                         I seek   16
This unfrequented place to find some *ease*,   17
Ease to the *body* some, *none* to the *mind*   18
From restless thoughts, that like a deadly *swarm*   19
Of hornets *armed*, no sooner found alone,   20
But rush upon me thronging, and *present*   21
Times *past, what once* I *was*, and *what am now*.   22

The hornets are really less important than such patternings as ease/ease; some...swarm...armed; present/past (a pun). To avoid being hypnotized by the mere patterns, though, keep checking them for local energy and variety (here, 18 and 22 are perhaps just thudding echoes of each other), and for consonance with the content (the rhyme system does buzz through the lines; ease/ease is like a settling down).

The word-order will often seem eccentric, following Latin practice rather than English; many of the words lean on their Latin sense; and there may also be an experimental torturing of the lines – try paraphrasing 1018–30. But the issue is, again, what that tortuousness contributes to the meaning – syntax agonistes. On the other hand, rhythm and diction are often colloquial – but easily mis-read because the context may be starchily latinate:

And expiate, if possible, my crime,
Shameful garrulity. To have revealed
Secrets of *men*, the secrets of a *friend*,
How heinous had the fact been – how deserving
Contempt and scorn of all, to be excluded
All friendship, and avoided as a BLAB,
The mark of fool set on his front!
But I *God's* counsel have not kept...   490

The only way to grasp the tortured energy and muted originality of the verse of *Samson* is to enact it, say it aloud, dwelling in its twists and turns. Don't expect imagery, or the shock of recognition; only an oral ritualizing of conflict, word by word, phrase by phrase representing sides and states in the struggle. Below is a model of lines 1335–40:

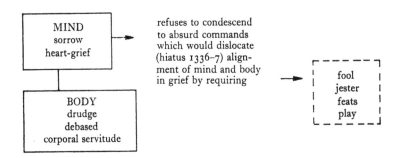

To take your reading one step further, think of forms that the refused command might take in life – to write? to be a husband?

There is violence in the verse; many of the words and phrases bang against each other irreconcilably – 'Effeminately vanquished' (562), 'trivial weapon' (263). The rhythms sometimes clench like teeth:

> The sun to me is dark
> And silent as the moon   86

Or they express sheer force thrust up and up:

> That heroic, that renowned,
> Irresistible Samson?   125

> Universally crowned with highest praises.   175

The Chorus is embarrassing at times:

> But what it is, hard is to say,
> Harder to hit,
> (Which way soever men refer it)   1013

This is because M took over bits of the style of the choruses in Greek tragedy, without transmuting them from one culture to another. Cranmer's Anglican litany will sound as silly if translated with equal naivety by a Greek in 4000 AD. But the Chorus are capable of both the tenderness and the exultation of lyric – 'This, this is he...softly a while' (115); 'When God into the hands of their deliverer *Puts* in*vi*ncible *might*' (1270).

When you are accustomed to the taste of the verse, look at local cases as models of the whole. Consider Samson's speech at 606–51. You might make notes of this sort:

139

Structure 3 strophës or stanzas, to that extent like a chorus (how does it compare, then, with a Shakespearean hero's soliloquy, or one of Browning's dramatic monologues?). But a deeper structure of 5 time phases – present future past present future. Tendency to circle, to get nowhere.

Content confirms that: second stanza repeats first, third repeats several other speeches, including his last at 590–8. But again a slightly deeper content, about internal pain; at its worst this is expressed in terms almost of self-torture, self-rape, where the torment of his griefs preys on *her* spirits – the mind's – and chews her up; chews himself up in 624 as if castrating himself – 'Mangle my apprehensive tenderest parts' – or putting his own eyes out.

Language and rhythm similarly: Words attack each other in echo and contradiction:

Left	sight	reserved	subject
helpless	alive	to be repeated	cruelty
loss			scorn

First strophe ends with weak flattening, bathos: 'but more intense' followed by the negative, obvious, otiose 'Though void of corporal sense'. It is as if the language commits suicide.

Conclusion. The speech is deeply 'about' the frustration and turning inward in self-enmity of aggression; hence contortion, disintegration, torture.

There is little point in trying to visualize the action on the mind's stage; but all the more in trying to hear the verse. Remember that variations in pace, and in pitch, and pauses, are far more effective than stress and volume. Because there is little interaction in *Samson* of the usual dramatic kind, it is also important to work into the central theme, the deep structure, kernel, gist – various words are used for it. 'The critic is trying to *synopsize* the given work...not in the degenerated sense... but in the sense of "conveying comprehensively", or "getting at the basis of"' (Kenneth Burke *Philosophy of literary form* 1957). Similarly, L. M. O'Toole defines kernel as

not the theme in the popular meaning of the word, some kind of 'digest' of the plot, but...a formulation in a more or less abstract 'meta-language' of the irreducible meaning of the text...a 'primary element' out of which the whole literary text is expanded...a combination of analysis and intuition enables one to make a provisional statement, a hypothesis. 'Analytic and synthetic approaches to narrative structure: Sherlock Holmes and *The Sussex vampire*' in *Style and structure in literature: essays in the new stylistics* ed. Roger Fowler, Blackwell, Oxford 1975

One of the most familiar ways of seeking the kernel is via clusters of associated words and, especially, images. *Samson* is awkward here: as the two essays on metaphor listed in the appendix say, the imagery is discordant. Still, it is up to you to either put it together, or consider what kind of 'kernel' it might be that lies inside such discordance. As one of them says:

Both loquacity and shipwreck are differing ways Samson has of speaking of separate but finally related aspects of his downfall...Samson's opening words, along with the first metaphor (to 114), call attention to several of the main ideas in the tragedy as well as to the main disorders which afflict and confuse him. To summarize by a brief analogy; *impotent* is to *disarm'd* as *servitude* is to *service* – fundamentally different but similar sounding contradictions...we readers are forced to make the unmade links between coin and swarm and head...batteries, unlock, grain, shook off snares, yoked, blot.

B. E. McCarthy 'Metaphor and plot in *SA*'
M quarterly VI no 4, Dec 1972

Talking

While you are learning how the verse goes, start considering some of the themes of the poem so far as they concern *you* (not M). It is important to do this by free discussion in a group of not more than about nine people, preferably mixed in age, gender and interests. In each case, let the discussion flow where it will (see Jane Abercrombie *Anatomy of judgement* 1960). Here are some themes; no doubt others will emerge from your reading:

vengeance	suicide	blindness	hair
patriotism	despair	strength/weakness	the mill

Writing and publication

We don't know when M wrote *SA*; so far, even statistical tests on the language have not helped. It was probably either c. 1648–50, when he was becoming completely blind, towards the end of the civil war; or after the Restoration of 1660. He published it together with *Paradise regained* in 1671, with a preface defending tragedy (even when not meant for the stage) against puritan censure. We print his preface, and the 'argument' or plot outline, at the end of this introduction.

Tragedy

The tragedy is here divided, for the reader's convenience, into the parts defined by Aristotle in his *Poetics* (section x or in some versions xii). They are: *prologos*; *parode* – the chorus's first ode; *episodes* – events, interaction; *stasima* – choral odes between the episodes; *exode* – final event, catastrophe; *kommos* – lament by chorus and remaining actors jointly. This is entirely for the reader's convenience and has no authority, though it is clear that Milton was following Aristotle's analysis – more closely than any actual Greek tragedy did. The Greeks produced tragedies in all-day trilogies, followed by an obscene satyr play, as part of the spring-time religious festival in honour of Dionysos, god of wine, fertility, the phallus, ecstacy, the bull and the goat. The festivals were conducted, at civic expense, in arenas holding perhaps 14,000 people in the open air. The actors (men only) wore high-heeled boots and large masks, and stayed on a low platform; while the chorus, of 15 with one flute-player, sang and danced in a 60-foot circle between the stage and the tiered seats of the audience. So a single tragedy was like a cross between an oratorio, a masked opera, and a very large ballet, on Horse Guards' Parade. Aristotle's analysis shows that the earliest trage-dies had been entirely choric; gradually dramatic episodes were inserted, until with some Greek dramatists they took over entirely.

A great deal of argument has been conducted about how 'Greek' M's poem is, and how 'tragic'; most of it seems pointless. The only thing you can safely say about tragedy in general is that it is the name given in western cultures to an enacted ritual contest between important powers in man, and society, and the universe, in which there is an element of grandeur and triumph, and an element of destruction and lament. Every culture has its own ritual and dramatic forms, its own relationship between art and religion, its own conceptions of what is great, fearsome, lamentable, of what the antagonisms of the world are. In M's case, we see that he rejected contemporary forms of both drama and liturgy (they had in any case both been prohibited in 1642);

that he produced a perfect version of an ancient Greek form, so going one better than that secular authority; that he dramatized a very primitive Hebrew legend, so civilizing, as it were, another ancestor, another authority; and that he prefaced his tragedy with a note that it 'never was intended' for the stage: the ritual in a sense then insulated from the test of enactment; the essentially public act of tragedy kept private; a way of maintaining the work's own authority inviolate? What does it mean to present, in a Christian culture, a tragedy (which in Greece would be dedicated to the god of wine) about the suicide (forbidden to Christians) of a Hebrew hero (the Jews were forbidden drama) who was not allowed to drink, or to cut his hair (Milton was not a Cavalier)?

Action of the drama, development of Samson

These are analysed in detail by Broadbent and Stein as cited in Resources p. 234 below. The way to go about it for yourself is to write down a framework of the divisions into episodes etc indicated in the text. Then specify each for yourself more distinctly; and draw on evidence of (a) the action (usually some kind of temptation); (b) the psychic movement, ups and downs, the 'feel' of the verse, emphatic passages (let yourself go here; leave it to others to modify your views); (c) verbal themes, recurring images etc. So you might for the episode with Manoa get something like this:

326-651 2nd episode   *Samson and Manoa*
(a) Samson and Manoa = sun and rest; but S refuses to set, to 'go gentle into that good night'. Manoa is his father, and an old man; so the temptation is to depend on parents, on the past, or on age and domesticity – anyway, to be dependent; specifically, to relinquish sex and striving.

(b) 566... fantasy of 'contemptible old age obscure' as another kind of mill, another kind of absurd spectacle; verbal disgust.

    590... nadir of poem: dark, droop, flat – blindness + depression + impotence.

    606... internal torture (but since it is internal, why is the language so physical?).

(c) 547... cool pure water versus wine/women (wet, hot, intoxicating) recurs at 625 cool herb/air versus inflamed mind...

Here are a few more notes on those categories:

(a) The action: oddly enough the 'temptations' fit the theory of group dynamics formulated by W. R. Bion in his *Experiences in groups* (1961). Bion says that members bring to a group three kinds of subconscious assumption which he labels dependence, pairing, fight/flight. These cause the members to behave as if they were there to depend on somebody else for help, as if they were there to reproduce themselves, or as if they were there to fight or run away from somebody – instead of actually getting on with the task of the group. You will have noticed these kinds of behaviour in committees, seminars and so on. The practical problem is how to mobilize those forces – dependence, pairing, fight/flight – in a sophisticated way, to get work done. If you now consider the tragedy's episodes in terms of these categories you may find some fit; and the Chorus emerges as at times a kind of facilitator, at others a group of which Samson is leader. The question remains, however, whether Samson's final act is (to use the jargon of group facilitators) task-directed? or a more elaborate evasion?

(b) To grapple with that question, watch out for the kinds or tones of dejection in each episode; and for the occasional promptings of refreshment, rousing motions and so on (eg 187, 665): are they valid?

(c) When tracing verbal themes, watch out for the items listed above for initial free discussion; and also for the following sets (you may not agree with the groupings): sting, snake, sore, balm etc; armour, weapons, snaies, traps etc; cool, air, water etc; exposure, fool, ridiculous. (See Carey and McCarthy in Resources p. 234 below.)

The Samson myth

The Samson story
The legends are told in *Judges* xiii–xvi. *Judges* is about the period *c.* 1200 BC when the Israelites, having escaped from bondage in Egypt and wandering in Sinai, had invaded Canaan and were settling it. They were still having to fight other settlers, especially Canaanites to the northwest and Philistines to the southwest; they seem to be largely under the power of the Philistines for

the Samson stories. The Israelites were still split into their twelve tribes, each with its own territory. They did not yet have a king, but local chieftains and guerrilla leaders called 'judges'; each judge tended to be a champion against a particular enemy. In the earliest versions of the story, Samson was not a 'judge' at all; but he came to be credited with the championship of Israel against the Philistines. (It was King David who finally defeated them: *II Samuel* viii.)

Uses in Christianity
During the later middle ages and the renaissance, Samson had been dignified by attention to his noble end rather than his earlier scandals. Several of his adventures were seen as foretypes of New Testament events: his taking away of the gates of Gaza was depicted as a type of Christ breaking out of the sepulchre to redeem man; his annunciation as parallel to Christ's.

For 17th-century English puritans, in addition, Samson and the other heroes of *Judges* – Jael, Gideon, Jephtha – offered a settler situation topical to the emigrations to America; and also to the conflict at home. These heroes could be seen as pure, natural republicans opposing corrupt city-dwelling monarchies. They are well-plotted and ferocious tales, too. Paul refers to several of them in *Hebrews* xi:

Now faith is the substance of things hoped for, the evidence of things not seen. For by it the elders [ie those before Christ's salvation] obtained a good report...By faith the walls of Jericho fell down, after they were compassed about seven days. By faith the harlot Rahab perished not with them that believed not, when she had received the spies with peace. And what shall I more say? for the time would fail me to tell of Gideon [*Judges* vi], and of Barak [iv], and of Samson, and of Japhtha [xi]...who through faith subdued kingdoms, wrought righteousness, obtained promises, stopped the mouths of lions, quenched the violence of fire, escaped the edge of the sword, out of weakness were made strong, waxed valiant in fight, turned to flight the armies of the aliens...

Philistines and Dagon
The Philistines had probably come from Crete. They were settled on the fertile coast north and south of Gaza – rich in corn, vineyards and olive-groves. They were a highly developed urban people, however, with superior architecture and sculpture, and skills in the working of iron and gold. They had unique political arrangements with 'lords' of some kind who could overrule both the kings of cities and the commanders of armies.

The lords are always referred to in the plural and in this they are like gods and giants (there were several Philistine giants, including Goliath whom David killed). On the other hand, like all the other inhabitants of Palestine, they worshipped gods other than Jahweh – notably Dagon. Marginal notes in the Bible identify Dagon with a fish-god from Babylon, whose idol was made like a merman; but the Hebrew word *dagan* means corn. The same doubt arises over the general name baalim for these baals, or gods, or lords, of the Palestinian tribes: were they gods of the earth (Babylonian Bel = earth and underworld) or gods of water? If they are gods of water, are they gods of natural watering, whether from streams in the earth, or rain from the sky? or gods of irrigation? The answer is, of course, that baalim in general, including Dagon, represent the essential twoness of experience, and especially of nature when men use it: the sea is defined by the land (the Philistines came from an island and lived along the coast); streams in the earth depend on rain from the sky; to make use of nature, man has to use artifice (cultivation, irrigation). In these ways Dagon was, you might say, a more realistic god than Jahweh, because more in accordance with the doubleness of experience and necessity; while Jahweh was essentially single, and demanded singleness of his people Israel.

The Israelites' dilemma: idolatry as whoredom
It is this demand that produced many of the agonized stories of *Joshua*, *Judges*, *Samuel* and *Kings*. If the settling Israelites stayed loyal to their god, they would be anomalies, living a 'puritan' nomad life learned in the desert, in another environment; but in order to share the prosperity of the other people living there, and learn agriculture, they would have to compromise with the alien fertility gods of the Canaanites and the Philistines – gods who did not live unseen in the sky but themselves compromised with the earth.

The dilemma is stated in *Hosea* ii: God argues against the Israelites as if they were his faithless wife; he claims it was he who gave her food and clothing, while she says it was her lovers – that is, the other inhabitants of Canaan; he threatens her with famine:

let her therefore put away her whoredoms out of her sight, and her adulteries from between her breasts; lest I strip her naked, and set her as in the day that she was born, and make her as a wilderness, and set her like a dry land, and slay her with thirst...For she said, 'I will go after my lovers, that give me my bread and my water, my wool and my flax, mine oil and my drink'...For she did not know that I gave her corn, and wine, and oil, and multiplied her silver and gold, wherewith they made Baal. Therefore will I return, and take away my corn in the time thereof, and my wine in the season thereof, and will take away my wool and my flax given to cover her nakedness...I will also cause all her mirth to cease, her feast days, her new moons, and her sabbaths, and all her solemn feasts. And I will destroy her vines and fig trees, whereof she hath said, 'These are my rewards that my lovers have given me': and I will make them a forest, and the beasts of the field shall eat them.

The prophets habitually used the language of adultery and whoredom for idolatry. It is the language of exchange: Israel sells sex for civilization. What it stands for, presumably, is the fact that cultures can learn from each other, and live in peace together, only by inter-marriage – by the exchange of sex, love, wives, children, as well as the exchange of food, money, cows, goods. On the other hand, when they do intermarry, they feel threatened by disloyalty, disunity.

We have therefore a dilemma of this kind set up:

religious taboo and tribal security demand endogamy (ie staying within the blood)	*but*	the survival of the tribe – food, goods, peace with neighbours – demands exogamy (ie recognizing affinal relations outside the blood)

Milton's dilemma

This dilemma is a familiar pattern in anthropology, but what relevance has this to Samson, and to M? Well of course M was caught in the same dilemma, as every intelligent revolutionary is; and, as a 17th-century Christian, he perceived the dilemma consciously as a version of the Israelite dilemma:

God	Satan
Israel as worshippers of Jahweh	gentiles
Israel as a political entity	Philistines
Christians	heathen
puritans	Roman catholics, Anglicans
parliamentarians	royalists

The two sides in the civil war actually insulted each other with these names (eg 'the bishops of England...Have they not been as the Canaanites and Philistines to this kingdom?' *Of reformation*; 'pure evangelic manna' versus 'tainted scraps' in *Of prelatical episcopacy*; the 'wondrous art Pontifical', ie of the pope, with which Sin and Death build their causeway from hell to earth *PL* x 312). It is easy for extremists on either side; but people who allow scope to their cieativity and complexity need something of both sides. M needed the heathen for they include Homer, Plato, Aeschylus, Virgil...He felt about some things in ways that needed images of royal courts, or catholic devotion, to express them:

And the apocalypse of St John is the majestic image of a high and stately tragedy, shutting up and intermingling her solemn scenes and acts with a sevenfold chorus of hallelujahs and harping symphonies...
Reason of church government

then with incredible expressions of joy all his brethren receive him [a prodigal], and set before him those perfumed banquets of Christian consolation: with precious ointments bathing and fomenting the old, and now to be forgotten, stripes which terror and shame had inflicted...
Ibid

Like Samson, M married the daughter of a royalist. His younger brother was a royalist. His nephew Edward Phillips was a rebel against the Commonwealth. His grandfather was a Roman catholic. He was a marginal man. He devoted much energy to trying to make sense of one side, which was a waste; but fundamentally Milton belongs to the race of marginal men, movers between worlds, inhabitants of shorelines – uneasy, incomplete, in conflict, indispensable.

Samson as mediator: anthropological analysis

Samson is more than marginal man: in the full anthropological sense he is a mediator (as the typologists had perceived when they equated him with Christ). That is why he was important to M, and why he may remain so to us.

In anthropological structures, a mediator is a being who has dual nationality and works across boundaries of existence, class, region. It is often anomalous, amphibious or fabulous: Christ, son of God and of man (Samson came in the middle ages to be regarded as a forerunner of Christ); the dolphins in Lycidas; angels. Now consider some of the facts about Samson.

Ancestry and birth
In the past, *Genesis* xxx shows that Samson's tribe, Dan, descended from *between* the two rival sisters Leah and Rachel from

whom sprang the rival kingdoms of Judah in the south and Israel in the north. Now Samson's mother is barren but an angel says she will bear a son: the meaning of this in myth is that the father is a god; so Samson is a demi-god.

Territory

Dan had two territories, one to the north, and one to the west of Jerusalem, which was Samson's. It lay on the north–south frontier and on the Israelite–Philistine border. It corresponded with the Shephéla or lowlands, strategic route from the south and the sea to the heart of Palestine. His village, and the villages of the woman of Timnath and Delilah, were in the Vale of Sorech: that means valley of the purple grape – a fertile vale joining Israel to Philistia. So Samson emerges as a frontiersman between the rich accomplished settlers of Philistia and the poor, half-wild immigrants of Israel.

Nature and culture

In anthropological structures these are binary oppositions which repeatedly have to be mediated. Nature includes the gods as well as wild animals, wild plants; culture is not artiness but civilization – cities, tools, cultivation, guilt and death. Here is a diagram of Samson in these terms:

nature = Samson	culture = Philistines
Kills with bare hands, natural weapons	Enslave him to a mill
Carries away loom, gates (probably same thing, primitive vertical loom): kills for clothes	Gates, machinery, textile industry
Breaks ropes as if with fire	Ropes = harness, domestic animals, pulleys, mills
Burns cultivated crops	Fertile land, developed agriculture
Must not drink wine but eats wild honey (mead? free sex?)	Feasts, commercial sex (prostitute of Gaza), legal marriage with bride-price etc (Timnath)
Destroys temple of Dagon, ie corn city	Destroyed by temple of Dagon

Sex, hair, eyes

The central part of Samson's activity has to do with women. He keeps seeking women across the frontier but he never establishes a marriage with one, and never has children; each tries to discover the secret of his strength and each time he takes violent revenge against their civilization; but when his hair is gone they are able to enslave him. He cannot manage sexual relations across the frontier; yet until that relationship is managed, Israel cannot learn from the Philistines nor the Philistines be safe from marauding. The sexual relationship is political: It is vital to mediate between the demand for pure endogamy, loyalty of blood and religion to Israel (which means isolation and poverty); and the need for exogamy, to learn from and live with the neighbouring tribes.

This is why his secret lies in his hair. The secret is sacredness, the fact that he is taboo, 'separate'. In the story he is taboo as a Nazarite; anthropologically, it is because he is a demi-god. Hair is universally regarded as a symbol of the sun and fertility; and of the genitals or genital fluids of either sex. Socially and in myth long hair is a sign that the wearer has *mana*, is royal etc (sun); and is sexually potent, and free.

This explains the blinding of Samson. The Philistines' attempts to civilize him are presented as various forms of bondage and finally as harnessing (as if he really were a wild animal) to a machine to grind their corn; so his strength is civilized by force. The symbol of this is the shaving of his hair, itself a symbol of castration but also of the sun being darkened; so his blinding is only another version of the same symbol for taming, for castration.

Milton and Samson

How poignant the myth of Samson was for M! At the personal level, was his blindness a psychosomatic castration? or was it the sign of inward illumination, of sanctity, as with Homer and Tiresias? How could he reconcile other oppositions in his life? He was a roundhead (a puritan tutor insisted that his hair be cropped at the age of ten); he was an enemy of the long-haired promiscuous royalists. Yet today he might have been called by the vulgar 'a long-haired leftie' – intellectual, radical, eccentric, violent to the point of regicide, like Samson in the flourish of his hairy strength. He persecuted urban culture, the frivolity of theatres, the debauchery of courtiers; yet he was of all men cultured, artificial in his art; he was a dramatist; and he not only married three times but publicly defended divorce and polygamy. What was he to make of his own contradictions? As Yeats said, out of the quarrel with others we make rhetoric;

out of the quarrel with ourselves, poetry. M's poetry is frequented with hairy half-gods – Comus, Satan like a comet, Adam with hyacinthine locks, the archangel Raphael whose six wings mantle his loins and shoulders 'with downy gold'...But there is no resolution. He makes a tragedy; but a tragedy is an expression, under ritual control, of forces and conflicts which off the stage, outside the temple, cannot be resolved.

Milton's preface

Of that sort of dramatic poem which is called tragedy

TRAGEDY, as it was anciently composed, hath been ever held the gravest, moralest, and most profitable of all other poems: therefore said by Aristotle to be of power by raising pity and fear, or terror, to purge the mind of those and such-like passions, that is to temper and reduce them to just measure with a kind of delight, stirred up by reading or seeing those passions well imitated. Nor is nature wanting in her own effects to make good his assertion: for so in physic things of melancholic hue and quality are used against melancholy, sour against sour, salt to remove salt humours. Hence philosophers and other gravest writers, as Cicero, Plutarch and others, frequently cite out of tragic poets, both to adorn and illustrate their discourse. The Apostle Paul himself thought it not unworthy to insert a verse of Euripides into the text of Holy Scripture, *I Cor* xv 33, and Paraeus commenting on the *Revelation*, divides the whole book as a tragedy, into acts distinguished each by a chorus of heavenly harpings and song between.

Heretofore men in highest dignity have laboured not a little to be thought able to compose a tragedy. Of that honour Dionysius the elder was no less ambitious, than before of his attaining to the tyranny. Augustus Caesar also had begun his *Ajax*, but unable to please his own judgement with what he had begun, left it unfinished. Seneca the philosopher is by some thought the author of those tragedies (at least the best of them) that go under that name. Gregory Nazianzen a Father of the Church, thought it not unbeseeming the sanctity of his person to write a tragedy, which he entitled, *Christ suffering*. This is mentioned to vindicate tragedy from the small esteem, or rather infamy, which in the account of many it undergoes at this day with other common interludes; happening through the poet's error of intermixing comic stuff with tragic sadness and gravity; or introducing trivial and vulgar persons, which by all judicious hath been counted absurd; and brought in without discretion, corruptly to gratify the people. And though ancient tragedy use no prologue, yet using sometimes in case of self-defence, or explanation, that which Martial calls an epistle; in behalf of this tragedy coming forth after the ancient manner, much different from what among us passes for best, thus much beforehand may be *epistled*; that chorus is here introduced after the Greek manner, not ancient only but modern, and still in use among the Italians. In the modelling therefore of this poem, with good reason, the ancients and Italians are rather followed, as of much more authority and fame.

The measure of verse used in the chorus is of all sorts, called by the Greeks monostrophic, or rather apolelymenon, without regard had to strophe, antistrophe or epode, which were a kind of stanzas framed only for the music, then used with the chorus that sung; not essential to the poem, and therefore not material; or being divided into stanzas or pauses, they may be called alloeostropha. Division into act and scene referring chiefly to the stage (to which this work never was intended) is here omitted.

It suffices if the whole drama be found not produced beyond the fifth act. Of the style and uniformity, and that commonly called the plot, whether intricate or explicit, which is nothing indeed but such economy, or disposition of the fable as may stand best with verisimilitude and decorum; they only will best judge who are not unacquainted with Aeschylus, Sophocles, and Euripides, the three tragic poets unequalled yet by any, and the best rule to all who endeavour to write tragedy. The circumscription of time wherein the whole drama begins and ends, is according to ancient rule, and best example, within the space of twenty-four hours.

The argument

Samson made captive, blind, and now in the prison at Gaza, there to labour as in a common workhouse, on a festival day, in the general cessation from labour, comes forth into the open air, to a place nigh, somewhat retired there to sit a while and bemoan his condition. Where he happens at length to be visited by certain friends and equals of his tribe, which make the Chorus, who seek to comfort him what they can; then by his old father Manoa, who endeavours the like, and withal tells him his purpose to procure his liberty by ransom; lastly, that this feast was proclaimed by the Philistines as a day of thanksgiving for their deliverance from the hands of Samson, which yet more troubles him. Manoa then departs to prosecute his endeavour with the Philistian lords for Samson's redemption: who in the meanwhile is visited by other persons; and lastly by a public officer to require his coming to the feast before the lords and people, to play or show his strength in their presence; he at first refuses, dismissing the public officer with absolute denial to come; at length persuaded inwardly that this was from God, he yields to go along with him, who came now the second time with great threatenings to fetch him; the Chorus yet remaining on the place, Manoa returns full of joyful hope, to procure ere long his son's deliverance: in the midst of which discourse an Hebrew comes in haste confusedly at first; and afterwards more distinctly relating the catastrophe, what Samson had done to the Philistines, and by accident to himself; wherewith the tragedy ends.

The persons

Samson, an Israelite of the tribe of Dan [his name means 'sun'. About 1600 the verb 'sampsoun' was coined to mean 'cast down in dejection and anguished thought']

Manoa, the father of Samson [means 'rest']

Dalila, Samson's second wife, a Philistine woman [in the middle ages, Dalida; later, Delilah; means, roughly, 'hang like a leaf' as in 727–8]

Harapha of Gath, a Philistine giant [his name means giant]

Officer of the Philistines

Messenger

Chorus of Danites [hence tribesmen and kinsmen of Samson and Manoa]

Agonistes is a Greek word which could mean the act of being an athlete; or a protagonist; or metaphorically being engaged in the *agon*, the game of life regarded as a spiritual struggle.

The scene before the prison in Gaza.

Samson Agonistes

Prologos

A LITTLE ONWARD lend thy guiding hand
    To these dark steps, a little further on;
    For yonder bank hath choice of sun or shade,
There I am wont to sit, when any chance
Relieves me from my task of servile toil,   5
Daily in the common prison else enjoined me,
Where I a prisoner chained, scarce freely draw
The air imprisoned also, close and damp,
Unwholesome draught: but here I feel amends,
The breath of heaven fresh blowing, pure and sweet,   10
With day-spring born; here leave me to respire.
  This day a solemn feast the people hold
To Dagon their sea-idol, and forbid
Laborious works, unwillingly this rest
Their superstition yields me; hence with leave   15
Retiring from the popular noise, I seek
This unfrequénted place to find some ease,
Ease to the body some, none to the mind
From restless thoughts, that like a deadly swarm

2 The line hesitates, shuffles, like blind footsteps. **6 else enjoined**
otherwise imposed on me. Note that 17c pronunciations would be servl,
tile, enjined. **9 draught** drawing in of the stale air. Starts series air,
cool etc. **11 dayspring** daybreak. Hints at salvation from the words of
the Benedictus celebrating the conception of John the Baptist; 'Whereby
the day-spring from on high hath visited us' *Luke* i. Also starts train of
references to special births. **respire** breathe. **12 people** the Philistines
Their feast parodies a Roman catholic holy day perhaps. **16 popular·**
people's – contemptuous, separating himself out.

Of hornets armed, no sooner found alone,   20
But rush upon me thronging, and present
Times past, what once I was, and what am now.
O wherefore was my birth from heaven foretold
Twice by an angel, who at last in sight
Of both my parents all in flames ascended   25
From off the altar, where an offering burned,
As in a fiery column charioting
His godlike presence, and from some great act
Or benefit revealed to Abraham's race?
Why was my breeding ordered and prescribed   30
As of a person separate to God,
Designed for great explóits; if I must die
Betrayed, captíved, and both my eyes put out,
Made of my enemies the scorn and gaze;
To grind in brazen fetters under task   35
With this heaven-gifted strength? O glorious strength
Put to the labour of a beast, debased
Lower than bond-slave! Promise was that I
Should Israel from Philístian yoke deliver;
Ask for this great deliverer now, and find him   40
Eyeless in Gaza at the mill with slaves,
Himself in bonds under Philístian yoke.
   Yet stay, let me not rashly call in doubt
Divine prediction; what if all foretold
Had been fulfilled but through mine own default,   45
Whom have I to complain of but myself?
Who, this high gift of strength committed to me,

20 armed first of a series of sting images. He means the hornet-thoughts
rush at him as soon as he's alone.   **28 and from** and flying up as if
from foretelling something great to the Hebrews.   **31 separate** dedicated,
singled out. The meaning of Nazarite.   **35 task** hard labour; piecework.
37 beast grinding was done by asses, and by slaves under punishment.
39 yoke subjugation, like an ox.   **41** Famous line, providing title of a
novel by Aldous Huxley. Eyeless in Gaza = blind among enemies like 68
which raises a question about blindness and its environments. Perhaps
also eyeless = loss of I, identity?   **43 stay** stop.   **45 Had** strongly stressed.
but except for.   **default** failure.   **47 Who** himself, having had the gift
committed to him, and knowing it lay in his hair and was easily taken
away etc.

In what part lodged, how easily bereft me,
Under the seal of silence could not keep
But weakly to a woman must reveal it,   50
O'ercome with importunity and tears.
O impotence of mind, in body strong!
But what is strength without a double share
Of wisdom – vast, unwieldy, burdensome,
Proudly secure, yet liable to fall   55
By weakest subtleties, not made to rule,
But to subserve where wisdom bears command.
God, when he gave me strength, to show withal
How slight the gift was, hung it in my hair.
    But peace, I must not quarrel with the will   60
Of highest dispensation, which herein
Haply had ends above my reach to know:
Suffices that to me strength is my bane,
And proves the source of all my miseries;
So many, and so huge, that each apart   65
Would ask a life to wail, but chief of all,
O loss of sight, of thee I most complain!
Blind among enemies, O worse than chains,
Dungeon, or beggary, or decrepit age!
Light the prime work of God to me is extinct,   70
And all her various objects of delight
Annulled, which might in part my grief have eased,
Inferior to the vilest now become
Of man or worm; the vilest here excel me,

48 part hair – or displaced from another part? Note emphasis of similar
long words in 51–2; cf 94, 119, 143, 194. 55 secure careless (Latin). 57
subserve strength meant to be subordinate to wisdom, not rule itself.
60 peace be quiet. 61 dispensation providence. 62 Haply perhaps.
ends intentions; hinting at the destruction of the Philistines at the end of
the poem, and also at Samson = Christ perhaps. 63 Suffices it's enough
that. ME and BANE are heavily stressed. 65 apart separately, would
need a lifetime to bewail. 69 Dungeon perhaps the imprisonment,
poverty and weakness here are not just Samson's. Are they dreadful?
is blindness among enemies worse? 70 prime the first act of creation
was when God said Fiat lux. extinct extinguished. 73 Inferior he is
now at the bottom of the scale of nature for here, in sight (which was
the senior sense), others are better.

They creep, yet see, I dark in light exposed   75
To daily fraud, contempt, abuse and wrong,
Within doors, or without, still as a fool,
In power of others, never in my own;
Scarce half I seem to live, dead more than half.
O dark, dark, dark, amid the blaze of noon,   80
Irrecoverably dark, total eclipse
Without all hope of day!
O first-created beam, and thou great word,
'Let there be light', and light was over all:
Why am I thus bereaved thy prime decree?   85
The sun to me is dark
And silent as the moon,
When she deserts the night
Hid in her vacant interlunar cave.
Since light so necessary is to life,   90
And almost life itself, if it be true
That light is in the soul,
She all in every part; why was the sight
To such a tender ball as the eye confined?
So obvious and so easy to be quenched,   95
And not as feeling through all parts diffused,
That she might look at will through every pore?
Then had I not been thus exiled from light;
As in the land of darkness yet in light,
To live a life half dead, a living death,   100
And buried; but O yet more miserable!
Myself, my sepulchre, a moving grave,

75 exposed part of a series on his public ridicule.   **77 still as a fool**
always treated like one of the idiots kept as clowns.   **80** Provides the
title of Arthur Koestler's novel *Darkness at noon*, and a line in Eliot's
East Coker.   **83 beam** both light, the first thing to be created; and (as
at *PL* iii 1) the Word, the Son who creates it.   **87 silent** used for the
part of the month when the moon doesn't shine but is on 'vacation',
supposedly in some hollow of the sky: but the ordinary senses of silent
and vacant also work.   **90 life** note the repetitions, rhyme etc here.
93 all the soul diffused throughout the body (like feeling in 96).   **95**
obvious exposed.   **97 she** the soul.   **99 land** ie death.   **102 Myself**
phonetically = miserable = sepulchre.

Buried, yet not exempt
By privilege of death and burial
From worst of other evils, pains and wrongs,   105
But made hereby obnoxious more
To all the miseries of life,
Life in captivity
Among inhuman foes.
     But who are these? for with joint pace I hear   110
The tread of many feet steering this way;
Perhaps my enemies who come to stare
At my affliction, and perhaps to insult,
Their daily practice to afflict me more.

CHORUS   *Parode*

This, this is he; softly a while,   115
Let us not break in upon him;
O change beyond report, thought, or belief!
See how he lies at random, carelessly diffused,
With languished head unpropped,
As one past hope, abandoned,   120
And by himself given over;
In slavish habit, ill-fitted weeds
O'er-worn and soiled.
Or do my eyes misrepresent? Can this be he,
That heroic, that renowned,   125
Irresistible Samson? whom unarmed
No strength of man, or fiercest wild beast could withstand;
Who tore the lion, as the lion tears the kid,
Ran on embattled armies clad in iron,
And weaponless himself,   130

106 obnoxious exposed. **110 joint** in step. Rhymes 110–112–114.
112 enemies along with affliction, worth looking up in a Bible concordance: eg *Psalm* xli. But whom he hears as enemies, quieten their
approach as friends. **119 languished** drooping; flower imagery here.
122 habit...weeds clothes. **124 my eyes** choruses speak collectively
and as individuals. Here collective hushing gives way to individual
questioning of sight, perhaps, in the imagination, a touching of their
eyes; and the end of 124 questions the beginning of 115.

Made arms ridiculous, useless the forgery
Of brazen shield and spear, the hammered cuirass,
Chalýbean-tempered steel, and frock of mail
Adamantean proof;
But safest he who stood aloof,   135
When insupportably his foot advanced,
In scorn of their proud arms and warlike tools,
Spurned them to death by troops. The bold Ascalonite
Fled from his lion ramp, old warriors turned
Their plated backs under his heel;   140
Or grovelling soiled their crested helmets in the dust.
Then with what trivial weapon came to hand,
The jaw of a dead ass, his sword of bone,
A thousand foreskins fell, the flower of Palestine
In Ramath-lechi famous to this day: 145
Then by main force pulled up, and on his shoulders bore
The gates of Azza, post, and massy bar
Up to the hill by Hebron, seat of giants old,
No journey of a sabbath-day, and loaded so;
Like whom the Gentiles feign to bear up heaven.   150
    Which shall I first bewail,
Thy bondage or lost sight,
Prison within prison
Inseparably dark?
Thou art become (O worst imprisonment!)   155
The dungeon of thyself; thy soul
(Which men enjoying sight oft without cause complain)
Imprisoned now indeed,
In reäl darkness of the body dwells,

131 **forgery** false strength of forged metal. 133 **Chalýbean** the Chalybes
were famous metalworkers of the ancient world. **frock** used of tunics and
coats for men till recently. 134 **Adamantean proof** armour (proof)
hard as a fabulous rock. 139 **old** veteran. 144 **foreskins** Philistines,
who were not circumcised (all Hebrews must be); a common OT usage
but 'sword of bone' imputes special phallic force to S here. 147 **Azza**
Gaza. 148 **giants** the Anakim of Hebron *Numbers* xiii, *Joshua* xiv. This
exploit used as a type of Christ's resurrection. Gaza to Hebron is about 40
miles; the Hebrews were not allowed to travel more than ¾ mile on the
sabbath. 150 **whom** Atlas. 157 **complain** it was a cliché that the
soul is a prisoner in the body.

Shut up from outward light 160
To incorporate with gloomy night;
For inward light alas
Puts forth no visual beam.
O mirror of our fickle state,
Since man on earth unparalleled! 165
The rarer thy example stands,
By how much from the top of wondrous glory,
Strongest of mortal men,
To lowest pitch of abject fortune thou art fallen.
For him I reckon not in high estate 170
Whom long descent of birth
Or the sphere of fortune raises;
But thee whose strength, while virtue was her mate,
Might have subdued the earth,
Universally crowned with highest praises. 175

SAMSON   *1st Episode*

I hear the sound of words, their sense the air
Dissolves unjointed ere it reach my ear.

CHORUS

He speaks, let us draw nigh. Matchless in might,
The glory late of Israel, now the grief;
We come thy friends and neighbours not unknown 180
From Eshtaol and Zora's fruitful vale

161 incorporate not clear whether the body is one body with night,
or the soul one body with the body. Clinking rhyme perhaps in the
way? **163 visual beam** physical light. **164 state** condition. It is the
mirror, S as 'model, that is unparalleled since the creation of man.
Your example is all the rarer for the depth of your fall. Individuals
commonly taken as examples or mirrors or lessons, especially of the
fickleness of fortune; and this kind of fall the basis of medieval and
some 16c tragedy; but what does mirror do to the visual imagery above?
170 estate rank. Tragic heroes were supposed to be of high rank; at
173 the Chorus justify S – 'But *thee*' – in spite of his not being so.
These lines are taken up again about 1660 ff. Note organ-like variety of
stresses in 175 contrasting with cadence of 169. **176** Onomatopoeia.
177 unjointed inarticulate. **179 late** recently. **181 Zora** S born there
and buried between it and Eshtaol: towns of the tribe of Dan.

To visit or bewail thee, or if better,
Counsel or consolation we may bring,
Salve to thy sores, apt words have power to 'suage
The tumours of a troubled mind,   185
And are as balm to festered wounds.

SAMSON

Your coming, friends, revives me, for I learn
Now of my own experience, not by talk,
How counterfeit a coin they are who friends
Bear in their superscription (of the most   190
I would be understood): in prosperous days
They swarm, but in adverse withdraw their head
Not to be found, though sought. Ye see, O friends,
How many evils have enclosed me round;
Yet that which was the worst now least afflicts me,   195
Blindness, for had I sight, confused with shame,
How could I once look up, or heave the head,
Who like a foolish pilot have shipwrecked
My vessel trusted to me from above,
Gloriously rigged; and for a word, a tear,   200
Fool, have divulged the secret gift of God
To a deceitful woman. Tell me, friends,
Am I not sung and proverbed for a fool
In every street, do they not say, how well
Are come upon him his deserts? Yet why?   205
Immeasurable strength they might behold
In me, of wisdom nothing more than mean;
This with the other should, at least, have paired,
These two proportioned ill drove me transverse.

184 sores part of a train of disease images. **190 superscription**
titling on a coin. Cf forgery 131? **of the most** understand that I
mean most, not all. **192 swarm** cf sting imagery. **198 shipwrecked**
beginning of a series; cf entry of Delilah. Rigged suggests his
hair and his potency. See *L. c.* **207 mean** average. **209 transverse**
maritime again.

CHORUS

Tax not divine disposal, wisest men   210
Have erred, and by bad women been deceived;
And shall again, pretend they ne'er so wise.
Deject not then so overmuch thyself,
Who hast of sorrow thy full load besides;
Yet truth to say, I oft have heard men wonder   215
Why thou shouldst wed Philístian women rather
Than of thine own tribe fairer, or as fair,
At least of thy own nation, and as noble.

SAMSON

The first I saw at Timna, and she pleased
Me, not my parents, that I sought to wed,   200
The daughter of an infidel: they knew not
That what I motioned was of God; I knew
From intimate impulse, and therefore urged
The marriage on; that by occasion hence
I might begin Israel's deliverance,   225
The work to which I was divinely called;
She proving false, the next I took to wife
(O that I never had! fond wish too late)
Was in the vale of Sorec, Dálila,
That specious monster, my accomplished snare.   230
I thought it lawful from my former act,
And the same end; still watching to oppress
Israel's oppressors: of what now I suffer
She was not the prime cause, but I myself,

212 **pretend** however wise they claim to be. 216 **Philístian** S had
first married a Philistine, the woman of Timnath (*Judges* xiv): then
visited a prostitute in Gaza: then taken Delilah as his mistress. 222
motioned S claims that the marriage he proposed was prompted by
God in a secret impulse, so he could use the occasion to work for
Israel. 230 **specious** deceptively attractive. **accomplished** well-
endowed, clever. 231 **lawful...end** says he married D; this would be
bigamy so excuses it by the treachery of first wife; and claims it was
again to infiltrate the enemy.

Who vanquished with a peal of words (O weakness!)   235
Gave up my fort of silence to a woman.

CHORUS

In seeking just occasion to provoke
The Philistine, thy country's enemy,
Thou never wast remiss, I bear thee witness:
Yet Israel still serves with all his sons.   240

SAMSON

That fault I take not on me, but transfer
On Israel's governors, and heads of tribes,
Who seeing those great acts which God had done
Singly by me against their conquerors
Acknowledged not, or not at all considered   245
Deliverance offered: I on the other side
Used no ambition to commend my deeds,
The deeds themselves, though mute, spoke loud the doer;
But they persisted deaf, and would not seem
To count them things worth notice, till at length   250
Their lords the Philistines with gathered powers
Entered Judéa seeking me, who then
Safe to the rock of Etham was retired,
Not flying, but forecasting in what place
To set upon them, what advantaged best;   255
Meanwhile the men of Judah to prevent
The harass of their land, beset me round;
I willingly on some conditions came
Into their hands, and they as gladly yield me
To the uncircumcised a welcome prey,   260
Bound with two cords; but cords to me were threads
Touched with the flame: on their whole host I flew
Unarmed, and with a trivial weapon felled
Their choicest youth; they only lived who fled.
Had Judah that day joined, or one whole tribe,   265
They had by this possessed the towers of Gath,

235 **peal** barrage. Elaborates the 'seal of silence' theme from 49; see the rest of this speech. 247 **ambition** influence. 266 **this** now.

And lorded over them whom now they serve;
But what more oft in nations grown corrupt,
And by their vices brought to servitude,
Than to love bondage more than liberty,　　　　　　270
Bondage with ease than strenuous liberty;
And to despise, or envy, or suspect
Whom God hath of his special favour raised
As their deliverer; if he aught begin,
How frequent to desert him, and at last　　　　　　275
To heap ingratitude on worthiest deeds?

CHORUS
Thy words to my remembrance bring
How Succoth and the fort of Penuel
Their great deliverer contemned,
The matchless Gideon in pursuit　　　　　　280
Of Madian and her vanquished kings;
And how ingrateful Ephraim
Had dealt with Jephtha, who by argument,
Not worse than by his shield and spear
Defended Israel from the Ammonite,　　　　　　285
Had not his prowess quelled their pride
In that sore battle when so many died
Without reprieve adjudged to death,
For want of well pronouncing *Shibboleth*.

SAMSON
Of such examples add me to the roll,　　　　　　290
Me easily indeed mine may neglect,
But God's proposed deliverance not so.

CHORUS　　　　　　　　　　　　　　　　*1st Stasimon*
Just are the ways of God,

268 Famous political statement; cf *PL* 11 255.　280 Gideon...Jephtha
heroes of Israel whose efforts were hindered by faction among the tribes:
Judges viii, xi. In revenge Jeptha killed 42,000 Ephraimites, recogniz-
able by not being able to pronounce the word *shibboleth* (=swelling corn
or flooding stream). Note rhyme.　291 mine my people.

And justifiable to men;
Unless there be who think not God at all,   295
If any be, they walk obscure;
For of such doctrine never was there school,
But the heart of the fool,
And no man therein doctor but himself.
    Yet more there be who doubt his ways not just,   300
As to his own edicts, found contradicting,
Then give the reins to wandering thought,
Regardless of his glory's diminution;
Till by their own perplexities involved
They ravel more, still less resolved,   305
But never find self-satisfying solution.
    As if they would confine the interminable,
And tie him to his own prescript,
Who made our laws to bind us, not himself,
And hath full right to exempt   310
Whom so it pleases him by choice
From national obstriction, without taint
Of sin, or legal debt;
For with his own laws he can best dispense.
    He would not else who never wanted means,   315
Nor in respect of the enemy just cause
To set his people free,
Have prompted this heroic Nazarite,
Against his vow of strictest purity,
To seek in marriage that fallacious bride,   320
Unclean, unchaste.
    Down reason then, at least vain reasonings down,
Though reason here aver
That moral verdict quits her of unclean:
Unchaste was subsequent, her stain not his.   325

298 **fool** 'The fool hath said in his heart, There is no God' *Ps* xiv.
299 **doctor** teacher, authority.   312 **obstriction** The Mosaic law for-
bidding marriage to Gentiles.   319 **purity** Nazarites could marry, but
not marry Gentiles.   320 **fallacious** false. But they go on to admit that
the woman of Timnath did not sin against her own morality by marrying
S; though she was later unfaithful to him.

But see here comes thy reverend sire
With careful step, locks white as down,
Old Manoa: advise
Forthwith how thou ought'st to receive him.

SAMSON

Ay me, another inward grief awaked, 330
With mention of that name renews the assault.

MANOA

Brethren and men of Dan, for such ye seem,
Though in this uncouth place; if old respect,
As I suppose, towards your once gloried friend,
My son now captive, hither hath informed 335
Your younger feet, while mine cast back with age
Came lagging after; say if he be here.

CHORUS

As signal now in low dejected state,
As erst in highest, behold him where he lies.

MANOA

O miserable change! is this the man, 340
That invincible Samson, far renowned,
The dread of Israel's foes, who with a strength
Equivalent to angels' walked their streets,
None offering fight; who single combatant
Duelled their armies ranked in proud array, 345
Himself an army, now unequal match
To save himself against a coward armed
At one spear's length. O ever-failing trust
In mortal strength! and O what not in man

326 reverend sire revered father.   **328 advise** consider.   **333 uncouth**
alien.   **338 signal** notable.

Deceivable and vain! Nay what thing good   350
Prayed for, but often proves our woe, our bane?
I prayed for children, and thought barrenness
In wedlock a reproach; I gained a son,
And such a son as all men hailed me happy;
Who would be now a father in my stead?   355
O wherefore did God grant me my request,
And as a blessing with such pomp adorned?
Why are his gifts desirable, to tempt
Our earnest prayers, then given with solemn hand
As graces, draw a scorpion's tail behind?   360
For this did the angel twice descend? for this
Ordained thy nurture holy, as of a plant;
Select, and sacred, glorious for a while,
The miracle of men: then in an hour
Ensnared, assaulted, overcome, led bound,   365
Thy foes' derision, captive, poor, and blind
Into a dungeon thrust, to work with slaves?
Alas methinks whom God hath chosen once
To worthiest deeds, if he through frailty err,
He should not so o'erwhelm, and as a thrall   370
Subject him to so foul indignities,
Be it but for honour's sake of former deeds.

SAMSON

Appoint not heavenly disposition, father,
Nothing of all these evils hath befallen me
But justly; I myself have brought them on,   375
Sole author I, sole cause: if aught seem vile,
As vile hath been my folly, who have profaned
The mystery of God given me under pledge
Of vow, and have betrayed it to a woman,
A Canaanite, my faithless enemy.   380
This well I knew, nor was at all surprised,

350 **Deceivable** all human experience is deceptive. **361 angel** *Judges*
xiii. **364 miracle** marvel. **373 Appoint** accuse: cf 210. **376 vile** my
folly was as vile as my condition seems. **381 surprised** taken by surprise.

But warned by oft experience: did not she
Of Timna first betray me, and reveal
The secret wrested from me in her height
Of nuptial love professed, carrying it straight   385
To them who had corrupted her, my spies,
And rivals? In this other was there found
More faith? who also in her prime of love,
Spousal embraces, vitiated with gold,
Though offered only, by the scent conceived   390
Her spurious first-born; treason against me?
Thrice she assayed with flattering prayers and sighs,
And amorous reproaches to win from me
My capital secret, in what part my strength
Lay stored, in what part summed, that she might know: 395
Thrice I deluded her, and turned to sport
Her importunity, each time perceiving
How openly and with what impudence
She purposed to betray me, and (which was worse
Than undissembled hate) with what contempt   400
She sought to make me traitor to myself;
Yet the fourth time, when mustering all her wiles,
With blandished parleys, feminine assaults,
Tongue-batteries, she surceased not day nor night
To storm me over-watched, and wearied out.   405
At times when men seek most repose and rest,
I yielded, and unlocked her all my heart,
Who with a grain of manhood well resolved
Might easily have shook off all her snares:
But foul effeminacy held me yoked   410
Her bond-slave; O indignity, O blot
To honour and religion! servile mind
Rewarded well with servile punishment!
The base degree to which I now am fallen,
These rags, this grinding, is not yet so base   415

387 **other** Delilah.   394 **capital** chief; of the head; fatal.   395 **summed**
concentrated.   404 **Tongue-batteries** military imagery.   405 **over-
watched** kept awake too long.   408 **resolved** resolute.

As was my former servitude, ignoble,
Unmanly, ignominious, infamous,
True slavery, and that blindness worse than this,
That saw not how degenerately I served.

MANOA

I cannot praise thy marriage-choices, son,   420
Rather approved them not; but thou didst plead
Divine impulsion prompting how thou might'st
Find some occasion to infest our foes.
I state not that; this I am sure; our foes
Found soon occasion thereby to make thee   425
Their captive, and their triumph; thou the sooner
Temptation found'st, or over-potent charms
To violate the sacred trust of silence
Deposited within thee; which to have kept
Tacit, was in thy power; true; and thou bear'st   430
Enough, and more the burden of that fault;
Bitterly hast thou paid, and still art paying
That rigid score.   A worse thing yet remains,
This day the Philistines a popular feast
Here celebrate in Gaza; and proclaim   435
Great pomp, and sacrifice, and praises loud
To Dagon, as their god who hath delivered
Thee Samson bound and blind into their hands,
Them out of thine, who slew'st them many a slain.
So Dagon shall be magnified, and God,   440
Besides whom is no god, compared with idols,
Disglorified, blasphemed, and had in scorn
By the idolatrous rout amidst their wine;
Which to have come to pass by means of thee,
Samson, of all thy sufferings think the heaviest,   445

421 Rather...in fact I disapproved of them. **422 impulsion** as in 223.
424 state I won't argue with that. **425 thee** stressed – cf their...their...
thou. **441 compared** brought down to the level of. **443 snorting
rhythm.**

Of all reproach the most with shame that ever
Could have befall'n thee and thy father's house.

SAMSON

Father, I do acknowledge and confess
That I this honour, I this pomp have brought
To Dagon, and advanced his praises high   450
Among the heathen round; to God have brought
Dishonour, obloquy, and oped the mouths
Of idolists, and atheists; have brought scandal
To Israel, diffidence of God, and doubt
In feeble hearts, propense enough before   455
To waver, or fall off and join with idols;
Which is my chief affliction, shame and sorrow,
The anguish of my soul, that suffers not
Mine eye to harbour sleep, or thoughts to rest.
This only hope relieves me, that the strife   460
With me hath end; all the contést is now
'Twixt God and Dagon; Dagon hath presumed,
Me overthrown, to enter lists with God,
His deity comparing and preferring
Before the God of Abraham. He, be sure,   465
Will not connive, or linger, thus provoked,
But will arise and his great name assert:
Dagon must stoop, and shall ere long receive
Such a discomfit, as shall quite despoil him
Of all these boasted trophies won on me,   470
And with confusion blank his worshippers.

MANOA

With cause this hope relieves thee, and these words
I as a prophecy receive: for God,
Nothing more certain, will not long defer
To vindicate the glory of his name   475
Against all competition, nor will long
Endure it, doubtful whether God be Lord,

454 **diffidence** distrust. 466 **connive** ignore. 471 **blank** confound.

Or Dagon. But for thee what shall be done?
Thou must not in the meanwhile here forgot
Lie in this miserable loathsome plight   480
Neglected. I already have made way
To some Philistian lords, with whom to treat
About thy ransom: well they may by this
Have satisfied their utmost of revenge
By pains and slaveries, worse than death inflicted   485
On thee, who now no more canst do them harm.

SAMSON

Spare that proposal, father, spare the trouble
Of that solicitation; let me here,
As I deserve, pay on my punishment;
And expiate, if possible, my crime,   490
Shameful garrulity. To have revealed
Secrets of men, the secrets of a friend,
How heinous had the fact been, how deserving
Contempt, and scorn of all, to be excluded
All friendship, and avoided as a blab,   495
The mark of fool set on his front!
But I God's counsel have not kept, his holy secret
Presumptuously have published, impiously,
Weakly at least, and shamefully: a sin
That Gentiles in their parables condemn   500
To their abyss and horrid pains confined.

MANOA

Be penitent and for thy fault contríte,
But act not in thy own affliction, son,
Repent the sin, but if the punishment
Thou canst avoid, self-preservation bids;   505

482 **treat** negotiate. 483 **this** now. 489 **pay on** go on paying. 493
fact deed. 496 **front** forehead. 500 **parables** myths, such as Tantalus,
punished in hades for betraying the gods' secrets. 505 **bids** the proper
motive of self-preservation bids you avoid punishment if you can:
anticipates suicide motif, though execution means only the carrying out
of punishment, not death.

Or the execution leave to high disposal,
And let another hand, not thine, exact
The penal forfeit from thyself; perhaps
God will relent, and quit thee all his debt;
Who evermore approves and more accepts   510
(Best pleased with humble and filial submission)
Him who imploring mercy sues for life,
Than who self-rigorous chooses death as due;
Which argues over-just, and self-displeased
For self-offence, more than for God offended.   515
Reject not then what offered means, who knows
But God hath set before us, to return thee
Home to thy country and his sacred house,
Where thou may'st bring thy offerings, to avert
His further ire, with prayers and vows renewed.   520

SAMSON

His pardon I implore; but as for life,
To what end should I seek it? when in strength
All mortals I excelled, and great in hopes
With youthful courage and magnanimous thoughts
Of birth from heaven foretold and high exploits,   525
Full of divine instínct, after some proof
Of acts indeed heroic, far beyond
The sons of Anak, famous now and blazed,
Fearless of danger, like a petty god
I walked about admired of all and dreaded   530
On hostile ground, none daring my affront.
Then swollen with pride into the snare I fell
Of fair fallacious looks, venereal trains,
Softened with pleasure and voluptuous life;
At length to lay my head and hallowed pledge   535
Of all my strength in the lascivious lap

511 Elision probably thum bland. **526 instínct** impulse. **528 Anak**
giants. **blazed** trumpeted. **533 venereal trains** sexual snares. **535**
pledge his hair. 'She made him sleep upon her knees' *Judges* xvi.

173

Of a deceitful concubine who shore me
Like a tame wether, all my precious fleece,
Then turned me out ridiculous, despoiled,
Shaven, and disarmed among my enemies. 540

CHORUS

Desire of wine and all delicious drinks,
Which many a famous warrior overturns,
Thou couldst repress, nor did the dancing ruby
Sparkling, out-poured, the flavour, or the smell,
Or taste that cheers the heart of gods and men, 545
Allure thee from the cool crystalline stream.

SAMSON

Wherever fountain or fresh current flowed
Against the eastern ray, translucent, pure
With touch ethereal of heaven's fiery rod
I drank, from the clear milky juice allaying 550
Thirst, and refreshed; nor envied them the grape
Whose heads that turbulent liquor fills with fumes.

CHORUS

O madness, to think use of strongest wines
And strongest drinks our chief support of health,
When God with these forbidden made choice to rear 555
His mighty champion, strong above compare,
Whose drink was only from the liquid brook.

SAMSON

But what availed this temperance, not complete
Against another object more enticing?
What boots it at one gate to make defence, 560

537 **concubine** mistress or extra wife: but he means Delilah, whom he
said he married. **shore** sheared me as if I were a castrated ram. **543
ruby** wine. As a Nazarite, S was teetotal. **547 fountain** spring. **548
Against** water flowing towards the east was supposed to be life-giving;
here it is purified by the sun's fiery ray. **550 milky juice** liquor sweet
as milk. **557 liquid** transparent (Latin *liquidus*).

And at another to let in the foe
Effeminately vanquished? by which means,
Now blind, disheartened, shamed, dishonoured, quelled,
To what can I be useful, wherein serve
My nation, and the work from heaven imposed,   565
But to sit idle on the household hearth,
A burdenous drone; to visitants a gaze,
Or pitied object, these redundant locks
Robustious to no purpose clustering down,
Vain monument of strength; till length of years   570
And sedentary numbness craze my limbs
To a contemptible old age obscure.
Here rather let me drudge and earn my bread,
Till vermin or the draff of servile food
Consume me, and oft-invocated death   575
Hasten the welcome end of all my pains.

MANOA

Wilt thou then serve the Philistines with that gift
Which was expressly given thee to annoy them?
Better at home lie bed-rid, not only idle,
Inglorious, unemployed, with age outworn.   580
But God who caused a fountain at thy prayer
From the dry ground to spring, thy thirst to allay
After the brunt of battle, can as easy
Cause light again within thy eyes to spring,
Wherewith to serve him better than thou hast;   585
And I persuade me so; why else this strength
Miraculous yet remaining in those locks?
His might continues in thee not for naught,
Nor shall his wondrous gifts be frustrate thus.

SAMSON

All otherwise to me my thoughts portend,   590

567 **gaze** spectacle.   568 **redundant** flowing.   569 **Robustious** powerful.   571 **craze** weaken, crack, make brittle.   574 **draff** garbage (which
was used to feed slaves). Such food will eat him, unless the rats do.
578 **annoy** hurt.   581 **fountain** *Judges* xv.

That these dark orbs no more shall treat with light,
Nor the other light of life continue long,
But yield to double darkness nigh at hand:
So much I feel my genial spirits droop,
My hopes all flat, nature within me seems   595
In all her functions weary of herself;
My race of glory run, and race of shame,
And I shall shortly be with them that rest.

MANOA

Believe not these suggestions, which proceed
From anguish of the mind and humours black,   600
That mingle with your fancy. I however
Must not omit a father's timely care
To prosecute the means of thy deliverance
By ransom or how else: meanwhile be calm,
And healing words from these thy friends admit.   605

SAMSON

O that torment should not be confined
To the body's wounds and sores
With maladies innumerable
In heart, head, breast, and reins;
But must secret passage find   610
To the inmost mind,
There exercise all his fierce accidents,
And on her purest spirits prey,
As on entrails, joints, and limbs,
With answerable pains, but more intense,   615
Though void of corporal sense.
   My griefs not only pain me
As a lingering disease,

594 **genial spirits** vital powers. 600 **humours black** supposed
liquid in the body of which an excess caused melancholy. 603
prosecute go on with. Manoah ignores S's despair. 609 **breast**
chest, torso; but as with heart and head the usage is psychosomatic.
reins kidneys; or urino-genital organs. 612 **accidents** symptoms (a
medical term then). 615 **answerable** mental pains correspondingly
more intense than physical.

But finding no redress, ferment and rage,
Nor less than wounds immedicable   620
Rankle, and fester, and gangrene,
To black mortification.
Thoughts my tormentors armed with deadly stings
Mangle my apprehensive tenderest parts,
Exasperate, exulcerate, and raise   625
Dire inflammation which no cooling herb
Or méd'cinal liquor can assuage,
Nor breath of vernal air from snowy alp.
Sleep hath forsook and given me o'er
To death's benumbing opium as my only cure.   630
Thence faintings, swoonings of despair,
And sense of heaven's desertion.
        I was his nursling once and choice delight,
His destined from the womb,
Promised by heavenly message twice descending.   635
Under his special eye
Abstemious I grew up and thrived amain;
He led me on to mightiest deeds
Above the nerve of mortal arm
Against the uncircumcised, our enemies.   640
But now hath cast me off as never known,
And to those cruel enemies,
Whom I by his appointment had provoked,
Left me all helpless with the irreparable loss
Of sight, reserved alive to be repeated   645
The subject of their cruelty, or scorn.
Nor am I in the list of them that hope;
Hopeless are all my evils, all remédiless;
This one prayer yet remains, might I be heard –
No long petition, speedy death,   650
The close of all my miseries, and the balm.

622 mortification the dead tissue of gangrene. **624 apprehensive**
sensitive. **639 nerve** sinew, strength. **643 his appointment** God's
command. **644** This line can run to several more than 10 syllables.

Many are the sayings of the wise
In ancient and in modern books enrolled;
Extolling patience as the truest fortitude;
And to the bearing well of all calamities,   655
All chances incident to man's frail life
Consólatories writ
With studied argument, and much persuasion sought
Lenient of grief and anxious thought,
But with the afflicted in his pangs their sound   660
Little prevails, or rather seems a tune,
Harsh, and of dissonant mood from his complaint,
Unless he feel within
Some source of consolation from above;
Secret refreshings, that repair his strength,   665
And fainting spirits uphold.
    God of our fathers, what is man!
That thou towards him with hand so various,
Or might I say contrarious,
Temper'st thy providence through his short course,   670
Not evenly, as thou rul'st
The angelic orders and inferior creatures mute,
Irràtional and brute.
Nor do I name of men the common rout,
That wandering loose about   675
Grow up and perish, as the summer fly,
Heads without name no more remembered,
But such as thou hast solemnly elected,
With gifts and graces eminently adorned
To some great work, thy glory,   680
And people's safety, which in part they effect:

654 **patience** we have specialized it to waiting: as a religious virtue it is more 'To do without, take tosses, and obey' without complaining (Hopkins *Patience, hard thing*). See Chorus again at 1287. 658 **sought** probably means argument that is painstakingly sought after, and which seeks to soothe grief. 667 **what is man** 'What is man, that thou art mindful of him?' *Ps* viii, *Job* vii. 670 **Temper'st** regulate, allot. 677 **Heads** try visualizing.

Yet toward these thus dignified, thou oft
Amidst their height of noon,
Changest thy countenance, and thy hand with no regard
Of highest favours past   685
From thee on them, or them to thee of service.
Nor only dost degrade them, or remit
To life obscured, which were a fair dismission,
But throw'st them lower than thou didst exalt them high,
Unseemly falls in human eye,   690
Too grievous for the trespass or omission,
Oft leav'st them to the hostile sword
Of heathen and profane, their carcases
To dogs and fowls a prey, or else captíved:
Or to the unjust tribunals, under change of times,   695
And condemnation of the ingrateful multitude.
If these they 'scape, perhaps in poverty
With sickness and disease thou bow'st them down,
Painful diseases and deformed,
In crude old age;   700
Though not disordinate, yet causeless suffering
The punishment of dissolute days, in fine,
Just or unjust, alike seem miserable,
For oft alike, both come to evil end.
    So deal not with this once thy glorious champion,   705
The image of thy strength, and mighty minister.
What do I beg? how hast thou dealt already?
Behold him in this state calamitous, and turn
His labours, for thou canst, to peaceful end.

                                            3rd Episode
But who is this, what thing of sea or land?   710
Female of sex it seems,

682 dignified worthily elevated. **687 degrade** lower in rank. **688 obscured** out of public notice. **690 falls** these changes do not fit ordinary
human notions of what is fair or appropriate. **700 crude** prematurely
aged. **701 not disordinate** even though not dissolute, yet they are punished
as if they were. **702 fine** conclusion. **705 So deal not** don't treat him like
that. **706 minister** servant. The line implies comparison with Christ.
710 sea links Delilah to Dagon; and the question makes her a hybrid.

That so bedecked, ornate, and gay,
Comes this way sailing
Like a stately ship
Of Tarsus, bound for the isles 715
Of Javan or Gadire
With all her bravery on, and tackle trim,
Sails filled, and streamers waving,
Courted by all the winds that hold them play,
An amber scent of odorous perfúme 720
Her harbinger, a damsel train behind;
Some rich Philistian matron she may seem,
And now at nearer view, no other certain
Than Dálila thy wife.

SAMSON

My wife, my traitress, let her not come near me. 725

CHORUS

Yet on she moves, now stands and eyes thee fixed,
About to have spoke, but now, with head declined
Like a fair flower surcharged with dew, she weeps
And words addressed seem into tears dissolved,
Wetting the borders of her silken veil: 730
But now again she makes address to speak.

DALILA

With doubtful feet and wavering resolution
I came, still dreading thy displeasure, Samson,
Which to have merited, without excuse,
I cannot but acknowledge; yet if tears 735
May expiate (though the fact more evil drew
In the perverse event than I foresaw)
My penance hath not slackened, though my pardon
No way assured. But conjugal affection

715 **Tarsus** 'ships of Tarshish' is a biblical phrase. Here it is sailing to
the Ionian isles (supposedly founded by Javan) or the other way to Cadiz.
720 **amber** like ambergris. 721 **damsel train** ladies in waiting.
727 **declined** what Delilah's name means. 729 **addressed** prepared.

Prevailing over fear, and timorous doubt   740
Hath led me on desirous to behold
Once more thy face, and know of thy estate:
If aught in my ability may serve
To lighten what thou suffer'st, and appease
Thy mind with what amends is in my power,   745
Though late, yet in some part to recompense
My rash but more unfortunate misdeed.

SAMSON

Out, out hyaena; these are thy wonted arts,
And arts of every woman false like thee,
To break all faith, all vows, deceive, betray,   750
Then as repentant to submit, beseech,
And reconcilement move with feigned remorse,
Confess, and promise wonders in her change,
Not truly penitent, but chief to try
Her husband, how far urged his patience bears,   755
His virtue or weakness which way to assail:
Then with more cautious and instructed skill
Again transgresses, and again submits;
That wisest and best men full oft beguiled,
With goodness principled not to reject   760
The penitent, but ever to forgive,
Are drawn to wear out miserable days,
Entangled with a poisonous bosom snake,
If not by quick destruction soon cut off
As I by thee, to ages an example.   765

DALILA

Yet hear me Samson; not that I endeavour
To lessen or extenuate my offence,
But that on the other side, if it be weighed

748 **hyaena** supposed to lure victims by calling like a human voice; often depicted as a foul female. **arts** magic spells, or craftiness. 751 **as** as if, pretending to be. 754 **try** test. 756 **virtue** strength. 763 **snake** recurs in this episode. 766 Delilah's syntax and intonation flutter; her theme is weakness.

By itself, with aggravations not surcharged,
Or else with just allowance counterpoised,   770
I may, if possible, thy pardon find
The easier towards me, or thy hatred less.
First granting, as I do, it was a weakness
In me, but incident to all our sex,
Curiosity, inquisitive, ímportune   775
Of secrets, then with like infirmity
To publish them, both common female faults:
Was it not weakness also to make known
For importunity, that is for naught,
Wherein consisted all thy strength and safety?   780
To what I did thou show'dst me first the way.
But I to enemies revealed, and should not.
Nor shouldst thou have trusted that to woman's frailty:
Ere I to thee, thou to thyself wast cruel.
Let weakness then with weakness come to parle   785
So near related, or the same of kind,
Thine forgive mine; that men may censure thine
The gentler, if severely thou exact not
More strength from me, than in thyself was found.
And what if love, which thou interpret'st hate,   790
The jealousy of love, powerful of sway
In human hearts, nor less in mine towards thee,
Caused what I did? I saw thee mutable
Of fancy, feared lest one day thou wouldst leave me
As her at Timna, sought by all means therefore   795
How to endear, and hold thee to me firmest:
No better way I saw than by impórtuning
To learn thy secrets, get into my power
Thy key of strength and safety: thou wilt say,
Why then revealed? I was assured by those   800
Who tempted me, that nothing was designed
Against thee but safe custody, and hold:
That made for me, I knew that liberty

793 mutable Of fancy fickle in love.   **803 made for me** was to my advantage.

182

Would draw thee forth to perilous enterprises,
While I at home sat full of cares and fears,   805
Wailing thy absence in my widowed bed;
Here I should still enjoy thee day and night
Mine and love's prisoner, not the Philistines',
Whole to myself, unhazarded abroad,
Fearless at home of partners in my love.   810
These reasons in love's law have passed for good,
Though fond and reasonless to some perhaps;
And love hath oft, well meaning, wrought much woe,
Yet always pity or pardon hath obtained.
Be not unlike all others, not austere   815
As thou art strong, inflexible as steel.
If thou in strength all mortals dost exceed,
In uncompassionate anger do not so.

SAMSON

How cunningly the sorceress displays
Her own transgressions, to upbraid me mine!   820
That malice not repentance brought thee hither,
By this appears: I gave, thou say'st, the example,
I led the way; bitter reproach, but true,
I to myself was false ere thou to me:
Such pardon therefore as I give my folly,   825
Take to thy wicked deed: which when thou seest
Impartial, self-severe, inéxorable,
Thou wilt renounce thy seeking, and much rather
Confess it feigned. Weakness is thy excuse,
And I believe it, weakness to resist   830
Philístian gold: if weakness may excuse,
What murderer, what traitor, parricide,
Incestuous, sacrilegious, but may plead it?
All wickedness is weakness: that plea therefore
With God or man will gain thee no remission.   835
But love constrained thee; call it furious rage
To satisfy thy lust: love seeks to have love;

812 fond silly.   **836 love** the phrase quotes her sarcastically.

My love how couldst thou hope, who took'st the way
To raise in me inéxpiable hate,
Knowing, as needs I must, by thee betrayed?   840
In vain thou striv'st to cover shame with shame,
Or by evasions thy crime uncover'st more.

DALILA

Since thou determin'st weakness for no plea
In man or woman, though to thy own condemning,
Hear what assaults I had, what snares besides,   845
What sieges girt me round, ere I consented;
Which might have awed the best-resolved of men,
The constantest to have yielded without blame.
It was not gold, as to my charge thou lay'st,
That wrought with me: thou know'st the magistrates   850
And princes of my country came in person,
Solicited, commanded, threatened, urged,
Adjured by all the bonds of civil duty
And of religion, pressed how just it was,
How honourable, how glorious to entrap   855
A common enemy, who had destroyed
Such numbers of our nation: and the priest
Was not behind, but ever at my ear,
Preaching how meritorious with the gods
It would be to ensnare an irreligious   860
Dishonourer of Dagon: what had I
To oppose against such powerful arguments?
Only my love of thee held long debate;
And cómbated in silence all these reasons
With hard contést: at length that grounded maxim   865
So rife and celebrated in the mouths

840 **by thee** that I had been betrayed by you. **841 shame** she is trying to
hide her guilt by acknowledging it, and saying he's guilty too. Next line
difficult, seems to have parallel meaning. Both lines share metaphor of un-
dressing, with implication that nakedness itself is shame or crime – indi-
cating S's feelings about Delilah now: cf 902. **843 determin'st** since you
judge weakness to be no excuse. **858 not behind** not backward: priest
joins with politician in plotting betrayal of a husband. **865 grounded**
established, basic. **866 rife and celebrated** widely held and respected.

Of wisest men, that to the public good
Private respects must yield, with grave authority
Took full possession of me and prevailed;
Virtue, as I thought, truth, duty so enjoining.   870

SAMSON

I thought where all thy circling wiles would end;
In feigned religion, smooth hypocrisy.
But had thy love, still odiously pretended,
Been, as it ought, sincere, it would have taught thee
Far other reasonings, brought forth other deeds.   875
I before all the daughters of my tribe
And of my nation chose thee from among
My enemies, loved thee, as too well thou knew'st,
Too well, unbosomed all my secrets to thee,
Not out of levity, but overpowered   880
By thy request, who could deny thee nothing;
Yet now am judged an enemy. Why then
Didst thou at first receive me for thy husband?
Then, as since then, thy country's foe professed:
Being once a wife, for me thou wast to leave   885
Parents and country; nor was I their subject,
Nor under their protection but my own,
Thou mine, not theirs: if aught against my life
Thy country sought of thee, it sought unjustly,
Against the law of nature, law of nations,   890
No more thy country, but an ímpious crew
Of men conspiring to uphold their state
By worse than hostile deeds, violating the ends
For which our country is a name so dear;
Not therefore to be obeyed. But zeal moved thee;   895
To please thy gods thou didst it. Gods unable

868 **respects** considerations, advantages. A form of utilitarianism. **870**
Virtue...it seemed to me then that virtue, truth (of ethics) and duty
commanded her to agree to betray him. **884 professed** I was just as
publicly your country's foe then as later. **893 violating the ends**
we love our country because it protects and mirrors our family: so if it
attacks our family it stops being our country. **895 zeal** sarcastic quoting
of her again.

To acquit themselves and prosecute their foes
But by ungodly deeds, the contradiction
Of their own deity, gods cannot be;
Less therefore to be pleased, obeyed, or feared.   900
These false pretexts and varnished colours failing,
Bare in thy guilt how foul must thou appear!

DALILA

In argument with men a woman ever
Goes by the worse, whatever be her cause.

SAMSON

For want of words no doubt, or lack of breath,   905
Witness when I was worried with thy peals.

DALILA

I was a fool, too rash, and quite mistaken
In what I thought would have succeeded best.
Let me obtain forgiveness of thee, Samson,
Afford me place to show what recompense   910
Towards thee I intend for what I have misdone,
Misguided; only what remains past cure
Bear not too sensibly, nor still insist
To afflict thyself in vain: though sight be lost,
Life yet hath many solaces, enjoyed   915
Where other senses want not their delights
At home in leisure and domestic ease,
Exempt from many a care and chance to which
Eyesight exposes daily men abroad.
I to the lords will intercede, not doubting   920
Their favourable ear, that I may fetch thee
From forth this loathsome prison-house, to abide
With me, where my redoubled love and care
With nursing diligence, to me glad office,

904 **cause** legal case. 906 **worried** much stronger sense in 17c, still
to do with dogs harassing prey, especially at the throat. **peals** see
235. 910 **Afford me place** give me a chance. 913 **Bear**...don't
take it too hard. 916 **senses** where the senses other than sight
will not lack satisfaction. 919 **abroad** out of doors. 924 **office** duty.
186

May ever tend about thee to old age   925
With all things grateful cheered, and so supplied,
That what by me thou hast lost thou least shalt miss.

SAMSON

No, no, of my condition take no care;
It fits not; thou and I long since are twain;
Nor think me so unwary or accursed   930
To bring my feet again into the snare
Where once I have been caught; I know thy trains
Though dearly to my cost, thy gins, and toils;
Thy fair enchanted cup, and warbling charms
No more on me have power, their force is nulled,   935
So much of adder's wisdom I have learnt
To fence my ear against thy sorceries.
If in my flower of youth and strength, when all men
Loved, honoured, feared me, thou alone could hate me,
Thy husband, slight me, sell me, and forgo me;   940
How wouldst thou use me now, blind, and thereby
Deceivable, in most things as a child
Helpless, thence easily contemned, and scorned,
And last neglected? How wouldst thou insult
When I must live uxorious to thy will   945
In perfect thraldom, how again betray me,
Bearing my words and doings to the lords
To gloss upon, and censuring, frown or smile?
This jail I count the house of liberty
To thine whose doors my feet shall never enter.   950

DALILA

Let me approach at least, and touch thy hand.

926 **grateful** pleasant (Latin *gratus*).   929 **fits not** pointless, and unseemly.   932 **trains** snares, nets: so are **gins, toils**.   934 **cup** in Greek
mythology Circe was a witch who lured men to her island with singing
and gave them a magic drink which turned them into animals. She is
the mother of Comus in M's masque.   936 **adder's** adders supposed to
be deaf; and snakes to be wise.   942 **child Helpless** rhythm of 942, and
hiatus to 943, weak and staggering.   944 **insult** exult. 945 **uxorious** as
a husband submissive to his wife (Latin *uxor*).   948 **gloss** annotate.
951 A one-line speech. Action?

187

SAMSON

Not for thy life, lest fierce remembrance wake
My sudden rage to tear thee joint by joint.
At distance I forgive thee, go with that;
Bewail thy falsehood, and the pious works 955
It hath brought forth to make thee memorable
Among illustrious women, faithful wives:
Cherish thy hastened widowhood with the gold
Of matrimonial treason: so farewell.

DALILA

I see thou art implacable, more deaf 960
To prayers than winds and seas, yet winds to seas
Are reconciled at length, and sea to shore:
Thy anger, unappeasable, still rages,
Eternal tempest never to be calmed.
Why do I humble thus myself, and suing 965
For peace, reap nothing but repulse and hate?
Bid go with evil omen and the brand
Of infamy upon my name denounced?
To mix with thy concernments I desist
Henceforth, nor too much disapprove my own. 970
Fame if not double-faced is double-mouthed,
And with contráry blast proclaims most deeds,
On both his wings, one black, the other white,
Bears greatest names in his wild aery flight.
My name perhaps among the circumcised 975
In Dan, in Judah, and the bordering tribes,
To all posterity may stand defamed,
With malediction mentioned, and the blot
Of falsehood most uncónjugal traduced.
But in my country where I most desire, 980
In Ecron, Gaza, Asdod, and in Gath
I shall be named among the famousest

967 **Bid go** told to go with the nasty prophecy of 955–7, that she will become notorious, among good wives, for a bad one. 969 **concernments** interests.

Of women, sung at solemn festivals,
Living and dead recorded, who to save
Her country from a fierce destroyer, chose   985
Above the faith of wedlock-bands, my tomb
With odours visited and annual flowers.
Not less renowned than in Mount Ephraim
Jael, who with inhospitable guile
Smote Sisera sleeping through the temples nailed.   990
Nor shall I count it heinous to enjoy
The public marks of honour and reward
Conferred upon me for the piety
Which to my country I was judged to have shown.
At this whoever envies or repines,   995
I leave him to his lot, and like my own.

CHORUS

She's gone, a manifest serpent by her sting
Discovered in the end, till now concealed.

SAMSON

So let her go, God sent her to debase me,
And aggravate my folly who committed   1000
To such a viper his most sacred trust
Of secrecy, my safety, and my life.

CHORUS

Yet beauty, though injurious, hath strange power,
After offence returning, to regain

984 recorded while still alive, and after my death, celebrated as the
woman who put her country above her marriage vows. Her point is that
even ethical reputation is cultural. Cf 1733.   **987 odours** incense.   **989
Jael** the oldest poem in the Bible in *Judges* iv celebrates Jael, an Israelite
woman who gave refuge and hospitality (and, it is implied, love) to
Sisera, a general of the hostile Canaanite army. Then while he slept
she hammered a tent-peg through his head. The song exults at the
physical triumph of a woman over a strong man. Mt Ephraim was
where Deborah lived.   **993 piety** in Rome, where the word comes
from, it was close to patriotism.   **998 Discovered** revealed, in her final
speech: and sexual?

Love once possessed, nor can be easily   1005
Repulsed, without much inward passion felt
And secret sting of amorous remorse.

SAMSON

Love-quarrels oft in pleasing concord end,
Not wedlock-treachery endangering life.

CHORUS   *3rd Stasimon*

It is not virtue, wisdom, valour, wit,   1010
Strength, comeliness of shape, or amplest merit
That woman's love can win or long inherit;
But what it is, hard is to say,
Harder to hit,
(Which way soever men refer it)   1015
Much like thy riddle, Samson, in one day
Or seven, though one should musing sit.
If any of these or all, the Timnian bride
Had not so soon preferred
Thy paranymph, worthless to thee compared,   1020
Successor in thy bed,
Nor both so loosely disallied
Their nuptials, nor this last so treacherously
Had shorn the fatal harvest of thy head.
Is it for that such outward ornament   1025
Was lavished on their sex, that inward gifts
Were left for haste unfinished, judgement scant,
Capacity not raised to apprehend
Or value what is best

1006 **passion** suffering. 1012 **inherit** it is not those qualities in men
which win, or for long retain, the love of women. 1016 **riddle** 'Out
of the eater came forth meat, out of the strong came forth sweetness'
Judges xiv. 1018 **If any** had it been any of those qualities that
counted with women, your first wife, the woman of Timnath, would
not have gone off with your best man (*Judges* xiv). 1023 **this last**
Delilah. 1025 The argument about the relationship between women's
beauty and men's weakness, women's incompleteness and men's
strength, is elaborated by Adam and Raphael at *PL* viii 528–611.

In choice, but oftest to affect the wrong? 1030
Or was too much of self-love mixed,
Of constancy no root infixed,
That either they love nothing, or not long?
　Whate'er it be, to wisest men and best
Seeming at first all heavenly under virgin veil, 1035
Soft, modest, meek, demure,
Once joined, the contrary she proves, a thorn
Intestine, far within defensive arms
A cleaving mischief, in his way to virtue
Adverse and turbulent, or by her charms 1040
Draws him awry enslaved
With dotage, and his sense depraved
To folly and shameful deeds which ruin ends.
What pilot so expért but needs must wreck
Embarked with such a steers-mate at the helm? 1045
　Favoured of heaven who finds
One virtuous rarely found,
That in domestic good combines:
Happy that house! his way to peace is smooth:
But virtue which breaks through all opposition, 1050
And all temptation can remove,
Most shines and most is acceptable above.
　Therefore God's universal law
Gave to the man despotic power
Over his female in due awe, 1055
Nor from that right to part an hour,
Smile she or lour:
So shall he least confusion draw
On his whole life, not swayed
By female usurpation, nor dismayed. 1060

1030 affect prefer. **1037 joined** in marriage. **1038 Intestine** internal, sometimes used for civil war. **1039 cleaving** sticking. **1046 Favoured** he is favoured who finds one of the rarely-found virtuous wives. Quotes *Proverbs* xxxi, on which the next few lines about a virtuous woman are based. **1048 combines** more specific sense than now, of pairing – uniting, joining (com-) of two (bin-): in this case, to produce married good instead of married evil. **1050 But** states a preference on God's part for a more heroic virtue; 'breaks through' cf S at 1696. **1059 swayed** ruled.

But had we best retire, I see a storm?

SAMSON

Fair days have oft contracted wind and rain.

CHORUS

But this another kind of tempest brings.

SAMSON

Be less abstruse, my riddling days are past.

CHORUS

Look now for no enchanting voice, nor fear   1065
The bait of honeyed words; a rougher tongue
Draws hitherward, I know him by his stride,
The giant Harapha of Gath, his look
Haughty as is his pile high-built and proud.
Comes he in peace? what wind hath blown him hither   1070
I less conjecture than when first I saw
The sumptuous Dálila floating this way:
His habit carries peace, his brow defiance.

SAMSON

Or peace or not, alike to me he comes.

CHORUS

His fraught we soon shall know, he now arrives.   1075

HARAPHA

I come not Samson, to condole thy chance,
As these perhaps, yet wish it had not been,
Though for no friendly intent. I am of Gath,

1068 **Harapha** Hebrew Rephaim = giants: by a euphemism to ward off
evil, common in myth, it comes from *rapha* = weak.   1069 **pile** building.
1073 **habit** clothing: he is not in armour.   1074 **Or** whether.   1075
fraught cargo.   1077 **these** Chorus.

Men call me Harapha, of stock renowned
As Og or Anak and the Emims old   1080
That Kiriathaim held, thou know'st me now
If thou at all art known. Much I have heard
Of thy prodigious might and feats performed
Incredible to me, in this displeased,
That I was never present on the place   1085
Of those encounters, where we might have tried
Each other's force in camp or listed field:
And now am come to see of whom such noise
Hath walked about, and each limb to survey,
If thy appearance answer loud report.   1090

SAMSON

The way to know were not to see but taste.

HARAPHA

Dost thou already single me; I thought
Gyves and the mill had tamed thee? O that fortune
Had brought me to the field where thou art famed
To have wrought such wonders with an ass's jaw;   1095
I should have forced thee soon wish other arms,
Or left thy carcase where the ass lay thrown:
So had the glory of prowess been recovered
To Palestine, won by a Philistine
From the unforeskinned race, of whom thou bear'st   1100
The highest name for valiant acts, that honour
Certain to have won by mortal duel from thee,
I lose, prevented by thy eyes put out.

SAMSON

Boast not of what thou wouldst have done, but do
What then thou wouldst, thou seest it in thy hand.   1105

1080 Og...giants; *Genesis* xiv, *Deuteronomy* ii, iii, *Numbers* xiii. **1087 listed** fitted with barriers for jousting. **1092 single** challenge. **1093 Gyves** fetters. **1099 Palestine** used here (as still by Arabs) in opposition to Israel. **1100 unforeskinned** circumcised, ie Jewish. **1105 in thy hand** easy, just in front of you.

HARAPHA

To combat with a blind man I disdain,
And thou hast need much washing to be touched.

SAMSON

Such usage as your honourable lords
Afford me assassinated and betrayed,
Who durst not with their whole united powers   1110
In fight withstand me single and unarmed,
Nor in the house with chamber ambushes
Close-banded durst attack me, no not sleeping,
Till they had hired a woman with their gold
Breaking her marriage faith to circumvent me.   1115
Therefore without feigned shifts, let be assigned
Some narrow place enclosed, where sight may give thee,
Or rather flight, no great advantage on me;
Then put on all thy gorgeous arms, thy helmet
And brigandine of brass, thy broad habergeon,   1120
Vantbrace and greaves, and gauntlet, add thy spear
A weaver's beam, and seven-times-folded shield,
I only with an oaken staff will meet thee,
And raise such outcries on thy clattered iron,
Which long shall not withhold me from thy head,   1125
That in a little time while breath remains thee,
Thou oft shalt wish thyself at Gath to boast
Again in safety what thou wouldst have done
To Samson, but shalt never see Gath more.

HARAPHA

Thou durst not thus disparage glorious arms   1130

1109 assassinated struck down by treachery (instead of fair fight);
from Moslem hashish-eaters hired to murder leaders of the crusades.
1110 powers army. **1112 chamber** bedroom. **1113 Close-banded**
by secret plot. **1116 feigned shifts** excuses for avoiding him. **1120
brigandine** body armour of rings or plates. **habergeon** chain-mail cape.
1121 Vantbrace fore-arm armour. **greaves** leg armour. **1122 beam**
roller in a loom. The shaft of Goliath's spear 'was like a weaver's beam'
I Samuel xvii. **1130 arms** Harapha defends chivalric honour, while S
here stands for the simple strength of the popular guerrilla.

Which greatest heroes have in battle worn,
Their ornament and safety, had not spells
And black enchantments, some magician's art,
Armed thee or charmed thee strong, which thou from
 heaven
Feign'dst at thy birth was given thee in thy hair,   1135
Where strength can least abide, though all thy hairs
Were bristles ranged like those that ridge the back
Of chafed wild boars or ruffled porcupines.

SAMSON

I know no spells, use no forbidden arts;
My trust is in the living God who gave me   1140
At my nativity this strength, diffused
No less through all my sinews, joints and bones,
Than thine, while I preserved these locks unshorn,
The pledge of my unviolated vow.
For proof hereof, if Dagon be thy god,   1145
Go to his temple, invocate his aid
With solemnest devotion, spread before him
How highly it concerns his glory now
To frústrate and dissolve these magic spells,
Which I to be the power of Israel's God   1150
Avow, and challenge Dagon to the test,
Offering to combat thee his champion bold,
With the utmost of his godhead seconded:
Then thou shalt see, or rather to thy sorrow
Soon feel, whose God is strongest, thine or mine.   1155

HARAPHA

Presume not on thy God, whate'er he be,
Thee he regards not, owns not, hath cut off

1133 **black** it was difficult to include such a folk-hero as S in Christian
tradition: one commentator did say he used black magic. Here M
allows him to answer the charge. 1138 **chafed** angry. 1140 **God**
'we trust in the living God, who is the saviour of all men' *I Timothy*
iv. 1152 **combat** offer yourself to the fight as Dagon's champion,
helped by all his divine power. 1157 **owns not** disowns, disavows.

Quite from his people, and delivered up
Into thy enemies' hand, permitted them
To put out both thine eyes, and fettered send thee   1160
Into the common prison, there to grind
Among the slaves and asses thy comrádes,
As good for nothing else, no better service
With those thy boisterous locks, no worthy match
For valour to assail, nor by the sword   1165
Of noble warrior, so to stain his honour,
But by the barber's razor best subdued.

SAMSON

All these indignities, for such they are
From thine, these evils I deserve and more,
Acknowledge them from God inflicted on me   1170
Justly, yet despair not of his final pardon
Whose ear is ever open; and his eye
Gracious to readmit the suppliant;
In confidence whereof I once again
Defy thee to the trial of mortal fight,   1175
By combat to decide whose god is God,
Thine or whom I with Israel's sons adore.

HARAPHA

Fair honour that thou dost thy God, in trusting
He will accept thee to defend his cause,
A murderer, a revolter, and a robber.   1180

SAMSON

Tongue-doughty giant, how dost thou prove me these?

HARAPHA

Is not thy nation subject to our lords?
Their magistrates confessed it, when they took thee

1167 **barber's** hairdresser regarded as the opposite of military. 1169
From thine from your people. Indignity meant a slight on real worth;
S deserves it from God, but not from the Philistines, who are worthless
themselves. 1181 **doughty** brave.

As a league-breaker and delivered bound
Into our hands: for hadst thou not committed   1185
Notorious murder on those thirty men
At Ascalon, who never did thee harm,
Then like a robber stripp'dst them of their robes?
The Philistines, when thou hadst broke the league,
Went up with armèd powers thee only seeking,   1190
To others did no violence nor spoil.

SAMSON

Among the daughters of the Philistines
I chose a wife, which argued me no foe;
And in your city held my nuptial feast:
But your ill-meaning politician lords,   1195
Under pretence of bridal friends and guests,
Appointed to await me thirty spies,
Who threatening cruel death constrained the bride
To wring from me and tell to them my secret,
That solved the riddle which I had proposed.   1200
When I perceived all set on enmity,
As on my enemies, wherever chanced,
I used hostility, and took their spoil
To pay my underminers in their coin.
My nation was subjected to your lords.   1205
It was the force of conquest; force with force
Is well ejected when the conquered can.
But I a private person, whom my country
As a league-breaker gave up bound, presumed
Single rebellion and did hostile acts!   1210

1184 league treaty. 1186 murder S had staked 30 sets of clothing that
the 30 Philistine guests at his wedding to the woman of Timnath could
not solve his riddle (line 1016). They got his wife to tell them the
answer by threatening her; so he killed 30 Philistines of Ascalon and
paid his bet with their clothes. This is another folk-hero event difficult
to hold in a religious myth. The word robber originally meant a robe
thief because so much wealth was carried in the form of clothes. 1195
politician scheming. 1208 private person but they were shocked
when I, an ordinary citizen, conducted an individual rebellion. S is
being sarcastic and goes on to claim that he was not ordinary but special.

I was no private but a person raised
With strength sufficient and command from heaven
To free my country; if their servile minds
Me their deliverer sent would not receive,
But to their masters gave me up for nought,   1215
The unworthier they; whence to this day they serve.
I was to do my part from heaven assigned,
And had performed it if my known offence
Had not disabled me, not all your force.
  These shifts refuted, answer thy appellant   1220
Though by his blindness maimed for high attempts,
Who now defies thee thrice to single fight,
As a petty enterprise of small enforce.

HARAPHA

With thee a man condemned, a slave enrolled,
Due by the law to capital punishment?   1225
To fight with thee no man of arms will deign.

SAMSON

Cam'st thou for this, vain boaster, to survey me,
To déscant on my strength, and give thy verdict?
Come nearer, part not hence so slight informed;
But take good heed my hand survey not thee.   1230

HARAPHA

O Baal-zebub! can my ears unused
Hear these dishonours, and not render death?

SAMSON

No man withholds thee, nothing from thy hand
Fear I incurable; bring up thy van,
My heels are fettered, but my fist is free.   1235

1218 offence telling Delilah his secret. 1220 shifts excuses. apellant
challenger. This is the third time. 1231 Baal-zebub lord of the flies,
a Philistine god. unused not used to hearing such insults. 1232 render
pay back, return. 1234 van vanguard, front line.

HARAPHA

This insolence other kind of answer fits.

SAMSON

Go baffled coward, lest I run upon thee,
Though in these chains, bulk without spirit vast,
And with one buffet lay thy structure low,
Or swing thee in the air, then dash thee down   1240
To the hazard of thy brains and shattered sides.

HARAPHA

By Ashtaroth ere long thou shalt lament
These braveries in irons loaden on thee.

CHORUS

His giantship is gone somewhat crestfallen,
Stalking with less unconscionable strides,   1245
And lower looks, but in a sultry chafe.

SAMSON

I dread him not, nor all his giant-brood,
Though fame divulge him father of five sons
All of gigantic size, Goliah chief.

CHORUS

He will directly to the lords, I fear,   1250
And with malicious counsel stir them up
Some way or other yet further to afflict thee.

SAMSON

He must allege some cause, and offered fight
Will not dare mention, lest a question rise

1237 **baffled** disgraced (a term of chivalry). 1242 **Ashtaroth** moon
goddess. 1243 **braveries** boasts. 1245 **unconscionable** extreme. 1248
Though...divulge even though he is said to be. 1253 **offered fight**
the fact that I offered to fight him.

Whether he durst accept the offer or not,   1255
And that he durst not plain enough appeared.
Much more affliction than already felt
They cannot well impose, nor I sustain,
If they intend advantage of my labours,
The work of many hands, which earns my keeping   1260
With no small profit daily to my owners.
But come what will, my deadliest foe will prove
My speediest friend, by death to rid me hence,
The worst that he can give, to me the best.
Yet so it may fall out, because their end   1265
Is hate, not help to me, it may with mine
Draw their own ruin who attempt the deed.

CHORUS   *4th Stasimon*
O how comely it is and how reviving
To the spirits of just men long oppressed!
When God into the hands of their deliverer   1270
Puts invincible might
To quell the mighty of the earth, the oppressor,
The brute and boisterous force of violent men
Hardy and industrious to support
Tyrannic power, but raging to pursue   1275
The righteous and all such as honour truth;
He all their ammunition
And feats of war defeats
With plain heroic magnitude of mind
And celestial vigour armed,   1280
Their armouries and magazines contemns,
Renders them useless, while
With wingèd expedition
Swift as the lightning glance he executes

1268 **Chorus** some astonishing staccato rhythms with crushing emphases (eg FEATS of WAR DE FEATS). 1277 **ammunition And feats** military potential, and acts of war. 1281 **magazines** stores. 1283 **expedition** speed. Probably pronounced expeditïön: if so, ammunitïön too. The imagery recurs at 1695.

His errand on the wicked, who surprised   1285
Lose their defence distracted and amazed.
   But patience is more oft the exercise
Of saints, the trial of their fortitude,
Making them each his own deliverer,
And victor over all   1290
That tyranny or fortune can inflict.
Either of these is in thy lot,
Samson, with might endued
Above the sons of men; but sight bereaved
May chance to number thee with those   1295
Whom patience finally must crown.

                                                    5th Episode
This idol's day hath been to thee no day of rest,
Labouring thy mind
More than the working day thy hands,
And yet perhaps more trouble is behind;   1300
For I descry this way
Some other tending, in his hand
A sceptre or quaint staff he bears,
Comes on amain, speed in his look.
By his habit I discern him now   1305
A public officer, and now at hand.
His message will be short and voluble.

OFFICER
Hebrews, the prisoner Samson here I seek.

CHORUS
His manacles remark him, there he sits.

OFFICER
Samson, to thee our lords thus bid me say;   1310

1285 errand purpose.   **1300 behind** to come.   **1303 quaint** ornamental.
1307 voluble quick.   **1309 remark** mark him out.

This day to Dagon is a solemn feast,
With sacrifices, triumph, pomp, and games;
Thy strength they know surpassing human rate,
And now some public proof thereof require
To honour this great feast, and great assembly;   1315
Rise therefore with all speed and come along,
Where I will see thee heartened and fresh clad
To appear as fits before the illustrious lords.

SAMSON

Thou know'st I am an Hebrew, therefore tell them,
Our law forbids at their religious rites   1320
My presence; for that cause I cannot come.

OFFICER

This answer, be assured, will not content them.

SAMSON

Have they not sword-players, and every sort
Of gymnic artists, wrestlers, riders, runners,
Jugglers and dancers, antics, mummers, mimics,   1325
But they must pick me out with shackles tired,
And over-laboured at their public mill,
To make them sport with blind activity?
Do they not seek occasion of new quarrels
On my refusal to distress me more,   1330
Or make a game of my calamities?
Return the way thou cam'st, I will not come.

OFFICER

Regard thyself, this will offend them highly.

1312 **triumph** celebration of victory.   1317 **heartened** refreshed, perhaps fed.   1323 **sword-players** fencers.   1324 **gymnic artists** acrobats.
1325 **antics** clowns. M bases his idea of the feast on the public sports held
in England on Sundays and festivals, especially May Day, until the
Commonwealth. Puritans objected to them as occasions of drunkenness
and lechery – like theatres.   1328 **blind activity** this stunning phrase
seems to bear a wider meaning than it needs for the context.   1333
Regard thyself take care.

SAMSON

Myself? my conscience and internal peace.
Can they think me so broken, so debased   1335
With corporal servitude, that my mind ever
Will condescend to such absurd commands?
Although their drudge, to be their fool or jester,
And in my midst of sorrow and heart-grief
To show them feats, and play before their god,   1340
The worst of all indignities, yet on me
Joined with extreme contempt? I will not come.

OFFICER

My message was imposed on me with speed,
Brooks no delay: is this thy resolution?

SAMSON

So take it with what speed thy message needs.   1345

OFFICER

I am sorry what this stoutness will produce.

SAMSON

Perhaps thou shalt have cause to sorrow indeed.

CHORUS

Consider, Samson; matters now are strained
Up to the height, whether to hold or break;
He's gone, and who knows how he may report   1350
Thy words by adding fuel to the flame?
Expect another message more imperious,
More lordly thundering than thou well wilt bear.

SAMSON

Shall I abuse this consecrated gift
Of strength, again returning with my hair   1355
After my great transgression, so requite

1342 **Joined** imposed.   1344 **Brooks** allows.   1346 **stoutness** obstinacy.

Favour renewed, and add a greater sin
By prostituting holy things to idols;
A Nazarite in place abominable
Vaunting my strength in honour to their Dagon?   1360
Besides, how vile, contemptible, ridiculous,
What act more execrably unclean, profane?

CHORUS

Yet with this strength thou serv'st the Philistines,
Idolatrous, uncircumcised, unclean.

SAMSON

Not in their idol-worship, but by labour   1365
Honest and lawful to deserve my food
Of those who have me in their civil power.

CHORUS

Where the heart joins not, outward acts defile not.

SAMSON

Where outward force constrains, the sentence holds
But who constrains me to the temple of Dagon,   1370
Not dragging? the Philistian lords command.
Commands are no constraints. If I obey them,
I do it freely; venturing to displease
God for the fear of man, and man prefer,
Set God behind: which in his jealousy   1375
Shall never, unrepented, find forgiveness.
Yet that he may dispense with me or thee
Present in temples at idolatrous rites
For some important cause, thou need'st not doubt.

CHORUS

How thou wilt here come off surmounts my reach.   1380

1367 **civil** of the state (not religion).   1369 **sentence** the maxim of ethics
just stated by the Chorus.   1377 **dispense** even so God may give me
or you a dispensation to be in an unclean place for an important purpose
(as S is going to be).   1380 **How**...I don't know how you're going to
escape this one.

SAMSON
Be of good courage, I begin to feel
Some rousing motions in me which dispose
To something extraördinary my thoughts.
I with this messenger will go along,
Nothing to do, be sure, that may dishonour   1385
Our Law, or stain my vow of Nazarite.
If there be aught of presage in the mind,
This day will be remarkable in my life
By some great act, or of my days the last.

CHORUS
In time thou hast resolved, the man returns.   1390

OFFICER
Samson, this second message from our lords
To thee I am bid say. Art thou our slave,
Our captive, at the public mill our drudge,
And dar'st thou at our sending and command
Dispute thy coming? come without delay;   1395
Or we shall find such engines to assail
And hamper thee, as thou shalt come of force
Though thou wert firmlier fastened than a rock.

SAMSON
I could be well content to try their art,
Which to no few of them would prove pernicious.   1400
Yet knowing their advantages too many,
Because they shall not trail me through their streets
Like a wild beast, I am content to go.
Masters' commands come with a power resistless
To such as owe them absolute subjection;   1405
And for a life who will not change his purpose?

1381 courage 'Wait on the Lord; be of good courage, and he shall
strengthen thy heart' *Ps* xxvii.   **1396 engines** machines such as pulleys; or
instruments of torture.   **1399 art** ingenuity.   **1404** S pretends to submit.

(So mutable are all the ways of men)
Yet this be sure, in nothing to comply
Scandalous or forbidden in our Law.

OFFICER

I praise thy resolution, doff these links: 1410
By this compliance thou wilt win the lords
To favour, and perhaps to set thee free.

SAMSON

Brethren farewell, your company along
I will not wish, lest it perhaps offend them
To see me girt with friends; and how the sight   1415
Of me as of a common enemy,
So dreaded once, may now exasperate them
I know not. Lords are lordliest in their wine;
And the well-feasted priest then soonest fired
With zeal, if aught religion seem concerned: 1420
No less the people on their holy-days
Impetuous, insolent, unquenchable;
Happen what may, of me expect to hear
Nothing dishonourable, impure, unworthy
Our God, our Law, my nation, or myself,   1425
The last of me or no I cannot warrant.

CHORUS   *5th Stasimon*

Go, and the Holy One
Of Israel be thy guide
To what may serve his glory best, and spread his name
Great among the heathen round: 1430
Send thee the angel of thy birth, to stand
Fast by thy side, who from thy father's field
Rode up in flames after his message told
Of thy conception, and be now a shield
Of fire; that spirit that first rushed on thee   1435

1426 last I can't guarantee whether this will be the end of me or not.

206

In the camp of Dan
Be efficacious in thee now at need.
For never was from heaven imparted
Measure of strength so great to mortal seed,
As in thy wondrous actions hath been seen.   1440

Exode

But wherefore comes old Manoa in such haste
With youthful steps? much livelier than erewhile
He seems: supposing here to find his son,
Or of him bringing to us some glad news?

MANOA

Peace with you brethren; my inducement hither   1445
Was not at present here to find my son,
By order of the lords new parted hence
To come and play before them at their feast.
I heard all as I came, the city rings
And numbers thither flock, I had no will,   1450
Lest I should see him forced to things unseemly.
But that which moved my coming now, was chiefly
To give ye part with me what hope I have
With good success to work his liberty.

CHORUS

That hope would much rejoice us to partake   1455
With thee; say reverend sire, we thirst to hear.

MANOA

I have attempted one by one the lords,
Either at home, or through the high street passing,
With supplication prone and father's tears
To accept of ransom for my son their prisoner,   1460
Some much averse I found and wondrous harsh,

1450 no will to go and watch. **1457 attempted** tried to persuade.
1459 prone on his face.

Contemptuous, proud, set on revenge and spite;
That part most reverenced Dagon and his priests,
Others more moderate seeming, but their aim
Private reward, for which both god and state   1465
They easily would set to sale, a third
More generous far and civil, who confessed
They had enough revenged, having reduced
Their foe to misery beneath their fears,
The rest was magnanimity to remit,   1470
If some convenient ransom were proposed.
    What noise or shout was that? It tore the sky.

CHORUS

Doubtless the people shouting to behold
Their once great dread, captive, and blind before them,
Or at some proof of strength before them shown.   1475

MANOA

His ransom, if my whole inheritance
May compass it, shall willingly be paid
And numbered down: much rather I shall choose
To live the poorest in my tribe, than richest,
And he in that calamitous prison left.   1480
No, I am fixed not to part hence without him.
For his redemption all my patrimony,
If need be, I am ready to forgo
And quit: not wanting him, I shall want nothing.

CHORUS

Fathers are wont to lay up for their sons,   1485
Thou for thy son art bent to lay out all;
Sons wont to nurse their parents in old age,
Thou in old age car'st how to nurse thy son,
Made older than thy age through eyesight lost.

1482 patrimony inheritance. 1484 wanting lacking. 1485 lay up...
out save...spend.

MANOA

It shall be my delight to tend his eyes,   1490
And view him sitting in the house, ennobled
With all those high exploits by him achieved,
And on his shoulders waving down those locks,
That of a nation armed the strength contained:
And I persuade me God had not permitted   1495
His strength again to grow up with his hair
Garrisoned round about him like a camp
Of faithful soldiery, were not his purpose
To use him further yet in some great service,
Not to sit idle with so great a gift   1500
Useless, and thence ridiculous about him.
And since his strength with eyesight was not lost,
God will restore him eyesight to his strength.

CHORUS

Thy hopes are not ill founded nor seem vain
Of his delivery, and thy joy thereon   1505
Conceived, agreeable to a father's love;
In both which we, as next, participate.

MANOA

I know your friendly minds and – O what noise?
Mercy of heaven what hideous noise was that?
Horribly loud unlike the former shout.   1510

CHORUS

Noise call you it or universal groan
As if the whole inhabitation perished,
Blood, death, and deathful deeds are in that noise,
Ruin, destruction at the utmost point.

1506 agreeable to in accordance with. The dialogue justifies false hopes
and rejoicing.

MANOA

Of ruin indeed methought I heard the noise,   1515
O it continues, they have slain my son.

CHORUS

Thy son is rather slaying them, that outcry
From slaughter of one foe could not ascend.

MANOA

Some dismal accident it needs must be;
What shall we do, stay here or run and see?   1520

CHORUS

Best keep together here, lest running thither
We unawares run into danger's mouth.
This evil on the Philistines is fallen,
From whom could else a general cry be heard?
The sufferers then will scarce molest us here,   1525
From other hands we need not much to fear.
What if his eyesight (for to Israel's God
Nothing is hard) by miracle restored,
He now be dealing dole among his foes,
And over heaps of slaughtered walk his way?   1530

MANOA

That were a joy presumptuous to be thought.

CHORUS

Yet God hath wrought things as incredible
For his people of old; what hinders now?

MANOA

He can I know, but doubt to think he will;
Yet hope would fain subscribe, and tempts belief.   1535
A little stay will bring some notice hither.

1519 dismal happening on an evil day. accident disaster. 1524 general
from everyone. 1529 dole pain. 1532 Colloquial rhythms for several
lines. 1535 fain subscribe like to think so. 1536 notice news.

CHORUS

Of good or bad so great, of bad the sooner;
For evil news rides post, while good news baits.
And to our wish I see one hither speeding,
An Ebrew, as I guess, and of our tribe.   1540

MESSENGER

O whither shall I run, or which way fly
The sight of this so horrid spectacle
Which erst my eyes beheld and yet behold;
For dire imagination still pursues me.
But providence or instinct of nature seems,   1545
Or reason though disturbed, and scarce consulted
To have guided me aright, I know not how,
To thee first reverend Manoa, and to these
My countrymen, whom here I knew remaining,
As at some distance from the place of horror,   1550
So in the sad event too much concerned.

MANOA

The accident was loud, and here before thee
With rueful cry, yet what it was we hear not,
No preface needs, thou seest we long to know.

MESSENGER

It would burst forth, but I recover breath,   1555
And sense distract, to know well what I utter.

MANOA

Tell us the sum, the circumstance defer.

MESSENGER

Gaza yet stands, but all her sons are fallen,
All in a moment overwhelmed and fallen.

1538 post fast (as the mail). baits pauses to feed (the horses). The
proverb continues the colloquiality in the face of the catastrophe. 1550
As...So though remaining distant from the grim event, yet too closely
involved in it.

MANOA

Sad, but thou know'st to Israelites not saddest   1560
The desolation of a hostile city.

MESSENGER

Feed on that first, there may in grief be surfeit.

MANOA

Relate by whom.

MESSENGER

                    By Samson.

MANOA

                                That still lessens
The sorrow, and converts it nigh to joy.

MESSENGER

Ah Manoa I refrain, too suddenly   1565
To utter what will come at last too soon;
Lest evil tidings with too rude irruption
Hitting thy agèd ear should pierce too deep.

MANOA

Suspense in news is torture, speak them out.

MESSENGER

Then take the worst in brief, Samson is dead.   1570

MANOA

The worst indeed, O all my hopes defeated
To free him hence! but death who sets all free
Hath paid his ransom now and full discharge.
What windy joy this day had I conceived
Hopeful of his delivery, which now proves   1575

1562 surfeit there may be too much grief.   **1574 windy** false pregnancy,
flatulence instead of a conception.

212

Abortive as the first-born bloom of spring
Nipped with the lagging rear of winter's frost.
Yet ere I give the reins to grief, say first,
How died he? death to life is crown or shame.
All by him fell thou say'st, by whom fell he,   1580
What glorious hand gave Samson his death's wound?

MESSENGER
Unwounded of his enemies he fell.

MANOA
Wearied with slaughter then or how? explain.

MESSENGER
By his own hands.

MANOA
            Self-violence? what cause
Brought him so soon at variance with himself   1585
Among his foes?

MESSENGER
          Inevitable cause
At once both to destroy and be destroyed;
The edifice where all were met to see him
Upon their heads and on his own he pulled.

MANOA
O lastly over-strong against thyself!   1590
A dreadful way thou took'st to thy revenge.
More than enough we know; but while things yet
Are in confusion, give us if thou canst,
Eye-witness of what first or last was done,
Relation more particular and distinct.   1595

MESSENGER
Occasions drew me early to this city,

1596 Occasions business.

And as the gates I entered with sun-rise,
The morning trumpets festival proclaimed
Through each high street: little I had dispatched
When all abroad was rumoured that this day   1600
Samson should be brought forth to show the people
Proof of his mighty strength in feats and games;
I sorrowed at his captive state, but minded
Not to be absent at that spectacle.
The building was a spacious theätre,   1605
Half round on two main pillars vaulted high,
With seats where all the lords and each degree
Of sort, might sit in order to behold,
The other side was open, where the throng
On banks and scaffolds under sky might stand;   1610
I among these aloof obscurely stood.
   The feast and noon grew high, and sacrifice
Had filled their hearts with mirth, high cheer, and wine,
When to their sports they turned. Immediately
Was Samson as a public servant brought,   1615
In their state livery clad; before him pipes
And timbrels, on each side went armèd guards,
Both horse and foot before him and behind
Archers, and slingers, cataphracts and spears.
At sight of him the people with a shout   1620
Rifted the air clamouring their god with praise,
Who had made their dreadful enemy their thrall.
He patient but undaunted where they led him,
Came to the place, and what was set before him
Which without help of eye might be assayed,   1625

1605 theätre in *A relation of a journey* to the middle east, 1615,
George Sandys reported a visit to Gaza, where 'On the north-east
corner and summit of the hill, are the ruins of huge arches sunk in
the earth, and other foundations of a stately building...The Jews do
fable this place to have bin the theatre of Samson, pulled down on
the head of the Philistines.' **1608 sort** rank. **in order** according to
rank. **1610 banks and scaffolds** benches and platforms. M is being
highly circumstantial, with Roman and Elizabethan theatres in mind.
1617 timbrels tambourines. **1619 cataphracts** armoured cavalry.
spears spearsmen.

To heave, pull, draw, or break, he still performed
All with incredible, stupendious force,
None daring to appear antagonist.
  At length for intermission sake they led him
Between the pillars; he his guide requested   1630
(For so from such as nearer stood we heard)
As over-tired to let him lean a while
With both his arms on those two massy pillars
That to the archèd roof gave main support.
He unsuspicious led him; which when Samson   1635
Felt in his arms, with head a while inclined,
And eyes fast fixed he stood, as one who prayed,
Or some great matter in his mind revolved.
At last with head erect thus cried aloud,
'Hitherto, lords, what your commands imposed   1640
I have performed, as reason was, obeying,
Not without wonder or delight beheld.
Now of my own accord such other trial
I mean to show you of my strength, yet greater;
As with amaze shall strike all who behold.'   1645
This uttered, straining all his nerves he bowed,
As with the force of winds and waters pent,
When mountains tremble, those two massy pillars
With horrible convulsion to and fro
He tugged, he shook, till down they came and drew   1650
The whole roof after them, with burst of thunder
Upon the heads of all who sat beneath,
Lords, ladies, captains, counsellors, or priests,
Their choice nobility and flower, not only
Of this but each Philistian city round   1655
Met from all parts to solemnize this feast.
Samson with these immixed, inevitably
Pulled down the same destruction on himself;
The vulgar only 'scaped who stood without.

1659 vulgar ordinary people.

O dearly-bought revenge, yet glorious!   1660
Living or dying thou hast fulfilled
The work for which thou wast foretold
To Israel, and now li'st victorious
Among thy slain self-killed
Not willingly, but tangled in the fold   1665
Of dire necessity, whose law in death conjoined
Thee with thy slaughtered foes in number more
Than all thy life had slain before.

SEMICHORUS I

While their hearts were jocund and sublime,
Drunk with idolatry, drunk with wine   1670
And fat regorged of bulls and goats,
Chanting their idol, and preferring
Before our living dread who dwells
In Silo his bright sanctuary:
Among them he a spirit of frenzy sent,   1675
Who hurt their minds,
And urged them on with mad desire
To call in haste for their destroyer;
They only set on sport and play
Unweetingly impórtuned   1680
Their own destruction to come speedy upon them.
So fond are mortal men
Fallen into wrath divine,
As their own ruin on themselves to invite,

1666 **necessity** the inevitability of 1586, 1657; the excuse for S's suicide. 1669 **sublime** elated. Lifting rhythms mime the Philistines' exaltation, and exult over their fall.   1671 **regorged** either gorged with, stuffed full of; or they sicked it up and ate again (as at Roman feasts). The point is that they were both doubly drunk, and doubly overfed.   1672 **preferring** boasting Dagon's superiority to God, who is a living object of worship (not a dead idol). The ark, where God manifested himself, was at Shiloh. 1680 **Unweetingly** unknowingly.   1682 **fond** foolish.

Insensate left, or to sense reprobate,   1685
And with blindness internal struck.

SEMICHORUS 2
But he though blind of sight,
Despised and thought extinguished quite,
With inward eyes illuminated
His fiery virtue roused   1690
From under ashes into sudden flame,
And as an evening dragon came,
Assailant on the perchèd roosts,
And nests in order ranged
Of tame villatic fowl; but as an eagle   1695
His cloudless thunder bolted on their heads.
So virtue given for lost,
Depressed, and overthrown, as seemed,
Like that self-begotten bird
In the Arabian woods embost,   1700
That no second knows nor third,
And lay erewhile a holocaust,
From out her ashy womb now teemed,
Revives, reflourishes, then vigorous most
When most unactive deemed,   1705

1685 Insensate... bereft of their senses; or left with only a corrupted mind.
1686 struck no rhyme ! Now S is celebrated in the opposite terms. **1690**
virtue strength and integrity. **1691 ashes** dead fire = blind eyes; also
anticipates phoenix 1699. **1692 dragon** firedrake, ie comet (with hairy
tail), or meteor, shooting star, supposed to look like flying dragons. OT
prophets threaten hostile cities with dragons. Latin *draco* = snake.
Some interpret it as solar serpent, symbol of renewed spirit of Samson.
1695 villatic farmyard – hens opposed by a phoenix, or a dragon.
1696 bolted like a meteorite or the bolt of a crossbow coming unseen
out of the blue; or a thunderbolt. All these forces are of air and fire,
from above. **1699 bird** the mythical phoenix existed only one at a time,
neuter, without mate or eggs. It lived in the spice forests of Arabia
for 500 years at a time. Then it made itself a pyre of spice wood and
burnt itself to death. Out of the ashes rose the next phoenix. An emblem
of Christ. See Lawrence's poem on it as emblem of destruction and
renewal of personality. **1700 embost** enclosed in the woods. **1702 holo-**
caust sacrificial creature completely burnt up. **1703 teemed** generated.

And though her body die, her fame survives,
A secular bird ages of lives.

MANOA

Come, come, no time for lamentation now,
Nor much more cause, Samson hath quit himself
Like Samson, and heroicly hath finished   1710
A life heroic, on his enemies
Fully revenged, hath left them years of mourning,
And lamentation to the sons of Caphtor
Through all Philistian bounds. To Israel
Honour hath left, and freedom, let but them   1715
Find courage to lay hold on this occasion,
To himself and father's house eternal fame;
And which is best and happiest yet, all this
With God not parted from him, as was feared,
But favouring and assisting to the end.   1720
Nothing is here for tears, nothing to wail
Or knock the breast, no weakness, no contempt,
Dispraise, or blame, nothing but well and fair,
And what may quiet us in a death so noble.
    Let us go find the body where it lies   1725
Soaked in his enemies' blood, and from the stream
With lavers pure, and cleansing herbs wash off
The clotted gore. I with what speed the while
(Gaza is not in plight to say us nay)
Will send for all my kindred, all my friends   1730
To fetch him hence and solemnly attend
With silent obsequy and funeral train
Home to his father's house: there will I build him
A monument, and plant it round with shade
Of laurel ever green, and branching palm,   1735
With all his trophies hung, and acts enrolled

1707 **secular** living through ages (Latin *secula seculorum* = for ever and
ever). 1713 **Caphtor** origin of the Philistines. 1722 **knock** Jewish
expression of grief. 1729 **plight** Gaza – ie the Philistines – is not in any
shape to forbid us.

In copious legend, or sweet lyric song,
Thither shall all the valiant youth resort,
And from his memory inflame their breasts
To matchless valour, and adventures high: 1740
The virgins also shall on feastful days
Visit his tomb with flowers, only bewailing
His lot unfortunate in nuptial choice,
From whence captivity and loss of eyes.

CHORUS

All is best, though we oft doubt,   1745
What the unsearchable dispose
Of Highest Wisdom brings about,
And ever best found in the close.
Oft he seems to hide his face,
But unexpectedly returns   1750
And to his faithful champion hath in place
Bore witness gloriously; whence Gaza mourns
And all that band them to resist
His uncontrollable intent.
His servants he with new acquist   1755
Of true experience from this great event
With peace and consolation hath dismissed,
And calm of mind all passion spent.

1737 **legend** writing on the tomb. 1748 **close** end. 1751 **in place**
here and now, and in good time. 1755 **acquist** increase, acquiring.

Appendix to
Samson Agonistes

Tragedy

See F. L. Lucas *Greek drama for everyman* 1954 with translations
of several, and Raymond Williams *Modern tragedy* 1966 and
Drama in performance 1964.

MILTON Commonplace Book c. 1637

For what in all philosophy is more important or more exalted than a
tragedy rightly produced, what more useful for seeing at a single view
the events and changes of human life?

FRIEDRICH NIETZSCHE Note 1052 of 1888 trans. Walter Kaufmann in his
ed, *Portable Nietzsche* Viking 1954

Dionysus versus 'the Crucified One': there you have the contrast. It is
not martyrdom that constitutes the difference – only here it has two
different senses. Life itself, its eternal fruitfulness and recurrence,
involves agony, destruction, the will to annihilation. In the other case,
suffering – 'the Crucified One as the Innocent One' – is considered
an objection to this life, as the formula of its condemnation. Clearly, the
problem is that of the meaning of suffering: whether a Christian meaning
or a tragic meaning. In the first case, it is supposed to be the path to a
sacred existence; in the second case, *existence is considered sacred enough*
to justify even a tremendous amount of suffering. The tragic man
affirms even the harshest suffering: he is sufficiently strong, rich, and
deifying for this; the Christian negates even the happiest life on earth:
he is sufficiently weak, poor, and disinherited to suffer from life in any
form. The God on the cross is a curse on life, a pointer to seek redemption
from it; Dionysus cut to pieces is a *promise* of life: it is eternally reborn
and comes back from destruction.

Samson in the 17th century

MILTON from *The reason of church government*

I shall show briefly, ere I conclude, that the prelates, as they are to the subjects a calamity, so are they the greatest underminers and betrayers of the monarch, to whom they seem to be most favourable. I cannot better liken the state and person of a king than to that mighty Nazarite Samson; who being disciplined from his birth in the precepts and the practice of temperance and sobriety, without the strong drink of injurious and excessive desires, grows up to a noble strength and perfection with those his illustrious and sunny locks, the laws, waving and curling about his godlike shoulders. And while he keeps them about him undiminished and unshorn, he may with the jawbone of an ass, that is, with the word of his meanest officer, suppress and put to confusion thousands of those that rise against his just power. But laying down his head among the strumpet flatteries of prelates, while he sleeps and thinks no harm, they wickedly shaving off all those bright and weighty tresses of his law, and just prerogatives, which were his ornament and strength, deliver him over to indirect and violent counsels, which, as those Philistines, put out the fair and far-sighted eyes of his natural discerning, and make him grind in the prisonhouse of their sinister ends and practices upon him: till he, knowing this prelatical razor to have bereft him of his wonted might, nourish again his puissant hair, the golden beams of law and right; and they sternly shook, thunder with ruin upon the heads of those his evil counsellors, but not without great affliction to himself. This is the sum of their loyal service to kings...

Areopagitica

Methinks I see in my mind a noble and puissant nation rousing herself like a strong man after sleep and shaking her invincible locks: methinks I see her as an eagle mewing her mighty youth, and kindling her undazzled eyes at the full midday beam; purging and unscaling her long-abused sight at the fountain itself of heavenly radiance; while the whole noise of timorous and flocking birds, with those that flutter about, amazed at what she means...

FRANCIS QUARLES from *The history of Samson* 1631

> When lusty diet and the frolic cup
> Had roused and raised their quickened spirits up
> And brave triúmphing Bacchus had displayed
> His conquering colours in their cheeks, they said:
>   'Call Samson forth. He must not work today,
> 'Tis a boon feast, we'll give him leave to play.
> Does he grind bravely? does our mill-horse sweat?
> Let him lack nothing – what he wants in meat,
> Supply in lashes. He is strong and stout.
> And with his breath can drive the mill about.
> He works too hard, we fear: go down and free him

Say that his mistress, Délila, would see him.
The sight of him will take our whöres short!
Go fetch him, then, to make our honours sport.
Bid him provide some riddles; let him bring
Some songs of triumph: he that's blind may sing
With better boldness. Bid him never doubt
To please – what matter though his eyes be out?
'Tis no dishonour that he cannot see:
Tell him the god of love's as blind as he!'
    With that they brought poor Samson to the hall,
And as he passed he gropes to find the wall;
His pace was slow, his feet were lifted high –
Each tongue would taunt him, every scornful eye
Was filled with laughter. Some would cry aloud,
'He walks in state! His lordship is grown proud!'
Some bid his honour, 'Hail!' while others cast
Reproachful terms upon him as he passed.
Some would salute him fairly and embrace
His wounded sides, then spit upon his face.
Others would cry, 'For shame! forbear to abuse
The high and great redeemer of the Jews!'
Some jibe and flout him with their taunts and quips
Whilst others flirt him on the starting lips.
    With that, poor Samson, whose abundant grief
Not finding hopes of comfort or relief,
Resolved for patience. Turning round, he made
Some shift to feel his keeper out, and said:
    'Good sir, my painful labour in the mill
Hath made me bold (although against my will)
To crave some little rest; if you will please
To let the pillar but afford some ease
To my worn limbs, your mercy should relieve
A soul that has no more but thanks to give.'
    The keeper yielded. Now the hall was filled
With princes and their people that beheld
Abusèd Samson, whilst the roof retained
A leash of thousands more whose eyes were chained
To this sad object with a full delight
To see this flesh-and-blood-relenting sight.
With that, the prisoner turned himself and prayed,
So soft that none but heaven could hear, and said:
    'My God, my God, although my sins do cry
For greater vengeance, yet thy gracious eye
Is full of mercy. O remember now
The gentle promise and that sacred vow
Thou mad'st to faithful Abram and his seed.
O hear my wounded soul, that has less need
Of life than mercy! Let thy tender care
Make good thy plenteous promise now, and hear!
See how thy cursèd enemies prevail
Above my strength: behold how poor and frail

My native power is; and, wanting thee,
What is there, O what is there, Lord, in me?
Nor is it I that suffer: my desért
May challenge greater vengeance if thou wert
Extreme to punish; Lord, the wrong is thine;
The punishment is just, and only mine.
I am thy champion, Lord. It is not me
They strike at: through my sides they thrust at thee;
Against thy glory 'tis their malice lies;
They aimed at that when they put out these eyes.
Alas! their blood-bedabbled hands would fly
On thee wert thou but clothed in flesh as I.
Revenge thy wrongs, great God! O let thy hand
Redeem thy suffering honour, and this land!
Lend me thy power, renew my wasted strength
That I may fight thy battles; and at length
Rescue thy glory that my hands may do
That faithful service they were born unto.
Lend me thy power that I may restore
Thy loss, and I will never urge thee more.'
   Thus having ended, both his arms he laid
Upon the pillars of the hall, and said:
   'Thus with the Philistines I resign my breath
And let my God find glory in my death.'
   And, having spoke, his yielding body strained
Upon those marble pillars that sustained
The ponderous roof; they cracked; and with their fall
Down fell the battlements and roof and all
And with their ruin slaughtered at a blow
The whole assembly: they that were below
Received their sudden deaths from those that fell
From off the top, whilst none was left to tell
The horrid shrieks that filled the spacious hall,
Whose ruins were impartial and slew all.
They fell; and, with an unexpected blow,
Gave everyone his death, and burial too.

Hair, eyes, sex

On eyesight/hair and sexuality, consider the blinding of Oedipus,
Tiresias (*PL* III 36 and Tennyson), and Gloucester in *Lear*; the
equation sun: strength:: shaving: weakness for Satan at *PL* I 596;
M's sonnets on his blindness, and 'tender ball' at *Samson* 94.
Milton equates them in *Divorce* where Eros 'darts the direct
rays of his then most piercing eyesight' but, finding the wrong
mate, his erection collapses like a headdress – that is, like cut
hair: 'his arrows lose their golden heads, and shed their purple
feathers, his silken braids untwine, and slip their knots, and

that original and fiery virtue...goes out'. The phrase 'fiery virtue' recurs in *Samson*, along with several others, eg 'ammunition, bounds, bondage, defilement, secret power, smarting corrosive, vigour and spirit'; above all 'to grind in the mill of an undelighted and servile copulation' (chapter VI). That is clinched by Blake's 'dark Satanic mills' in the lyric called 'Jerusalem' which is actually the prologue to his epic *Milton*. In the poem properly called *Jerusalem* he elaborates the image of the mill as a rationalizing force which enslaves men to a rigid society and a rigid ethic:

> I turn my eyes to the schools and universities of Europe
> And there behold the loom of Locke whose woof rages dire,
> Washed by the water-wheels of Newton. Black the cloth
> In heavy wreaths folds over every nation; cruel works
> Of many wheels I view, wheel without wheel, with cogs tyrannic
> Moving with compulsion each other: not as those in Eden, which
> Wheel within wheel in freedom revolve, in harmony and peace.
>
> XV 14

Consider the use of *cog* these days. Ask why it is normally assumed that Blake meant actual cotton mills? What are our mills? Look at the mills of hell in Bosch's Vienna *Last judgement*, or in Piranesi. M merges Samson's mill with prison, and prison with his body.

Marcuse uggests a way of seeing how Samson's sexuality and symbolic castration, and his binding to the mill, relate to his final act:

Civilization is first of all progress in *work* – that is, work for the procurement and augmentation of the necessities of life. This work is normally without satisfaction in itself; to Freud it is unpleasurable, painful... 'For what motive would induce man to put his sexual energy to other uses if by any disposal of it he could obtain fully satisfying pleasure? He would never let go of his pleasure and would make no further progress' ['The most prevalent form of degradation in erotic life' *Collected papers* IV]...Thus the main sphere of civilization appears as a sphere of *sublimation*. But sublimation involves *desexualization*... Culture demands continuous sublimation; it thereby weakens Eros, the builder of culture. And desexualization, by weakening Eros, unbinds the destructive impulses ['death instinct']. Civilization is thus threatened by an instinctual de-fusion, in which the death instinct strives to gain ascendancy over the life instincts. Originating in renunciation and developing under progressive renunciation, civilization tends toward self-destruction.
  This argument runs too smooth to be true...not all work involves desexualization...Secondly, the inhibitions enforced by culture also affect...the derivatives of the death instinct, aggressiveness and the

destructive impulses. In this respect at least, cultural inhibitions would accrue to the strength of Eros... However, while the destructive impulses are thus being satisfied, such satisfaction cannot stabilize their energy in the service of Eros. Their destructive force must drive them beyond this servitude and sublimation, for their aim is, not matter, not nature, not any object, but life itself.

Eros and civilization Sphere 1956 ch 4

For the further study of hair see Pope's *Rape of the lock* as illustrated by Beardsley; Charles Berg *The unconscious significance of hair* 1951 as criticized by E. R. Leach in 'Magical hair' *Journal of the Royal Anthropological Inst* LXXXVIII 1958; hair and its cognates in a concordance to Milton, especially in *Lycidas* where girl's hair turns into the thread of life which 'the blind Fury [Fate] with the abhorred shears...slits'. As well as the Fates of Greek myth see Berenice, whose hair kept her husband safe; Nisus, whose daughter Scylla pulled out his fatal gold hair; and Pterelaos similarly (usually cited under his daughter's lover, Amphitryon).

Peter Abelard (1079–1142) wrote a *planctus* or lament for Samson in Latin; here is part of it translated by Peter Dronke in his *Poetic individuality in the middle ages: new departures in poetry 1000–1150* (OUP 1970):

...Virorum fortissimum	That mightiest of men
nuntiatum per angelum,	whom an angel heralded,
Nazareum inclitum,	the renowned Nazarite,
Israelis clipeum –	shield of Israel –
cuius cor vel saxeum	whose heart is so like stone
non fleat sic perditum?	it will not weep that thus he fell?
Quem primo Dalida	Dalila robbed him first
sacra cesarie,	of his hallowed hair,
hunc hostes postea	then his enemies
privarunt lumine.	robbed him of light.
Exhaustus viribus,	Drained of his strength,
orbatus oculis,	bereft of his eyes,
mole fit deditus	consigned to the mill
athleta nobilis.	is the noble champion.
Clausus carcere,	Incarcerated,
oculorumque lumine	the light of his eyes
iam privatus,	now plucked away,
quasi geminis	as if with double
ad molam sudans tenebris	darkness toiling at the mill
est opressus.	he is weighed down.
Ludos marcios	He ravages the limbs
plus exercere solitos	more used to exercise
frangit artus.	in sports of war.
Hos cibario	Keeping those limbs

vix sustentans edulio	barely alive with fodder
iumentorum,	of beasts of burden,
quod – et nimius	eating rarely –
labor hic et insolitus –	and even to eat an immense
sumit rarum,	unwonted struggle –
crebris stimulis	with repeated goading
agitatur ab emulis	he is driven by his adversaries
ut iumentum...	like a beast.

Abelard, a brilliant academic philosopher in Paris, was castrated by enemies because of his love for his pupil Heloise; she became a nun, he retired to the abbey of St Denis and, later, the abbeys of St Gildas and Cluny. The *planctus* presents Samson, along with Adam, David and Solomon, as an example of a man ruined by a woman.

M's allegory of Eros and Anteros:

Fourthly, Marriage is a covenant, the very being whereof consists not in a forced cohabitation, and counterfeit performance of duties, but in unfeigned love and peace. And of matrimonial love, no doubt but that was chiefly meant which by the ancient sages was thus parabled: that Love, if he be not twin-born, yet hath a brother wondrous like him, called Anteros; whom while he seeks all about, his chance is to meet with many false and feigning desires that wander singly up and down in his likeness; by them in their borrowed garb, Love, though not wholly blind, as poets wrong him, yet having but one eye, as being born an archer aiming, and that eye not the quickest in this dark region here below, which is not Love's proper sphere, partly out of the simplicity and credulity which is native to him, often deceived, embraces and consorts him with these obvious and suborned striplings, as if they were his mother's own sons; for so he thinks them, while they subtilly keep themselves most on his blind side. But after a while, as his manner is, when soaring up into the high tower of his Apogaeum, above the shadow of the earth, he darts out the direct rays of his then most piercing eyesight upon the impostures and trim disguises that were used with him, and discerns that this is not his genuine brother, as he imagined; he has no longer the power to hold fellowship with such a personated mate: for straight his arrows lose their golden heads, and shed their purple feathers, his silken braids untwine, and slip their knots, and that original and fiery virtue given him by fate all on a sudden goes out, and leaves him undeified and despoiled of all his force. Till, finding Anteros at last, he kindles and repairs the almost-faded ammunition of his deity by the reflection of a coequal and homogeneal fire...this is a deep and serious verity, shewing us that love in marriage cannot live nor subsist unless it be mutual; and where love cannot be, there can be left of wedlock nothing but the empty husk of an outside matrimony, as undelightful and unpleasing to God as any other kind of hypocrisy.

Doctrine and discipline of divorce

Structuralism

If you want to conduct a structural analysis, study *Lévi-Strauss* (Fontana Modern Masters 1970; pp. 65–75 are a good entry) and 'The legitimacy of Solomon' in *Genesis as myth and other essays* (1969), both by E. R. Leach, and then proceed as follows:

1. Establish in your mind the habit of binary division, especially along the scale nature (including gods)...culture (men in cities, tools etc; culture in the arty sense is not intended specifically).

2. Establish the habit of putting binary pairs together so as to point up their relationship in the form A:B::C:D. Note that likely categories for A, B, C etc are blood relatives, affinal relatives, species of being (gods, men, animals etc), kinds of food and cooking, use or non-use of the senses (eg taste, silence...).

3. Watch out for boundaries (forest-plain; seashore).

4. Write out the entire set of Samson stories, one incident at a time, one sentence per incident. Number each. Eg

> 1. Samson is born to a barren woman on the promise of an angel, against taboos on hair, wine, unclean food.
> 2. The angel goes up in flames.
> 3. The child is born and called 'sun'.
> 4. Samson seeks woman of Timnath, of an enemy tribe...

5. Collect all the incident-sentences into a few categories which recur, eg

A. enemy women	B. fire
4. Samson seeks woman of Timnath	2. Angel goes up in flames
n. Samson and prostitute of Gaza	3. Samson means sun

6. Try concentrating some of the categories into single sentences (eg A. Samson habitually goes with foreign women) and then putting them up against each other (but he is a champion of Israel).

7. Then the difficult part. Among the riddles which remain to be solved are (a) the water from the jawbone: has it a parallel? what does it mean? note that it is a kind of food derived from a dead wild thing. (b) Samson's silence, and interest in riddles. (c) birds: the rock of Etam means place of the birds of prey; the jawbone-spring means spring of the partridge. I have been very

much helped in this area by the Provost of King's – not only his publications but his unfailing intellectual generosity; however, he says he doesn't know what the silence and the birds signify.

This kind of discussion meets the Marcusian in the problem of mediating between the Third World and western industrialism. Must the undeveloped countries 'succumb...to a more or less terroristic system of primary accumulation' (Marcuse *One-dimensional man* 1964 ch 2) like Samson in the end – with the implication of mutually annihiliating vengeance? or might the indigenous non-industrial technology and society contribute to a new kind of civilization?

Delilah

She is part of the mythology of the witch, the female who seduces to bestialize (Circe, the mother of Comus), which is also, because it displaces the phallic human head and reduces to servitude, to castrate – as more obviously with Salome (Wilde, Beardsley) and Cleopatra (Shakespeare, Shaw). M's Samson responds as misogynist. Either way, discussion is probably more fruitful than study of the analogues, so long as it does not run away into contempt of the fear of women: the question is, why? what is it all about? what is the opposite myth, the opposite prejudice?

Giants

Delilah's name means droop; it is odd that Harapha's does too. The Hebrew for giant is *raphah*, which means to go limp, to be appalled. So the word for giants is either a description of the dread they cause; or a euphemism – the dreadful strong are called the weak limp people just as the Greek Furies were called Eumenides, the friendlies. Giant was always used in the plural, *rephaim*, which suggests a race, a people; and *rephaim* also means ghosts, which suggests that the giants were the ancestors, the dead, the people already there, the fearsome forebears. Samson's strength has giant implications but as a young giant he is a defender against the *rephaim*, a Jack-the-giantkiller. On giants as a special case of potency but also, often, of playfulness, consult really good reference books on the Babylonian Enkidu (also very hairy) and the Greek Hercules, Alcyoneus etc.

Suicide and despair

KENNETH BURKE from 'The imagery of killing' *Hudson review* 1 1948 repr in his *A grammar of motives* World Publishing Co. 1950 as 'The "use" of M's Samson'

The recurring stress upon the *reflexive* nature of Samson's act (the element of self-destruction in his way of slaying the enemy) can be a roundabout device for sanctioning suicide: yet Milton's religion strongly forbade suicide. Compelled by his misfortunes to live with his rage, gnawed by resentments that he could no longer release fully in outward contest, Milton found in Samson a figure ambivalently fit to symbolize both aggressive and inturning trends. Here too, though still more remotely, would be 'literature for use': the poetic re-enactment of Samson's role could give pretexts for admitting a motive which, if not so clothed or complicated, if confronted in its simplicity, would have been inadmissible. By dramatic subterfuge Milton could include what he would have had to exclude, if reduced to a conceptually analytic statement.

The dramatic terms provide a rich context that greatly modifies whatever modicum of suicide may be present in the motivational formula as a whole. But all such important modifications, or qualifications, are dropped when we reduce the complexity to one essential strand, slant, or 'gist', isolating this one reflexive element as the implicitly dominant motive, an all-pervasive generating principle. And considered from the symbolic point of view, is nothing other than the *imaging of a fall*, expressed roundabout in doctrinal or 'educational' terms. Within the limitations imposed by the nature of the book, the final proclaiming of this 'law', in strict analogy with the accelerated motion of falling bodies, in its way expresses but the same leap into the cosmic abyss that Matthew Arnold expresses through the suicide of Empedocles.

The range of images that can be used for concretizing the process of transformation is limited only by the imagination and ingenuity of poets. But the selective nature of existence favors some images above others – and high among them, naturally, is the imagery of Life and Death, with its variants of being born, being reborn, dying, killing, and being killed. Consider, now, the hypothetical case of a poet who would identify himself with some particular imagery of transformation selected from this order of terms, terms using the imagery of Life and Death. We can easily conceive of a poet who, wanting to symbolize the transformation of some evil trait within himself, writes a poem accordingly; and in this poem he might identify himself with a figure who, marked by this trait, takes his own life, thereby ritualistically transforming the trait. (That is, if the figure in the fiction possessed some outstanding vice, and slew himself as an act of judgment against this vice, such imagery of suicide could be a ritualistic means whereby the poet sought to purge his own self of this vice, or purified the vice by identifying it with the dignity of death.) Or another might symbolize this same transformation by imaginatively endowing some 'outward enemy' with the trait, and then imaginatively slaying that enemy. Or a third poet might identify himself with a figure who possessed that trait, and then might imagine an enemy who slew his poetic counterpart. The trait, whatever its stylistic transformation (magnification, purification, martyrdom, etc.), may not even be 'slain' by an 'alien' principle at all, so far as the original poet was concerned; the contest may most likely symbolize the pitting of one motivational principle against another where *both* principles are strongly characteristic of the poet personally. (Think, for instance, of the 'murderous' relation between the critical and poetic 'selves' of T. S. Eliot, as symbolized in his *Murder in the Cathedral*, and previously discussed in our *Attitudes Toward History*.) Similarly, if a principle were located in the figure of mother, father, child, tyrant, or king, and were ritually transformed under these guises, we should have respectively: matricide, patricide, infanticide, tyrannicide, or regicide. The Nazis, locating the *transformandum* in the whole Jewish people as their chosen vessel, gave us a 'scientific' variant: genocide. And the frequent psycho-analytic search for 'unconscious' desires to kill some member of the family, either through rivalry or through love frustrated and expressed in reverse, puts the emphasis at the wrong place. For the so-called 'desire to kill' a certain person is much more properly analysable as a desire to *transform the principle* which that person *represents*.

See the 'terrible sonnets' of Hopkins – *Carrion comfort, No worst, there is none, Patience, My own heart let me more have pity on*; and especially this one, of 1885:

> I wake and feel the fell of dark, not day.
> What hours, O what black hoürs we have spent
> This night! what sights you, heart, saw; ways you went!
> And more must, in yet longer light's delay.
>   With witness I speak this. But where I say
> Hours I mean years, mean life. And my lament
> Is cries countless, cries like dead letters sent
> To dearest him that lives alas! away.
>
>   I am gall, I am heartburn. God's most deep decree
> Bitter would have me taste: my taste was me;
> Bones built in me, flesh filled, blood brimmed the curse.
>   Selfyeast of spirit a dull dough sours. I see
> The lost are like this, and their scourge to be
> As I am mine, their sweating selves; but worse.

ALBERT CAMUS 'Absurdity and suicide' in his *The myth of Sisyphus and other essays* Paris 1942 trans. Justin O'Brien, Hamilton 1955

There is but one truly serious philosophical problem and that is suicide. Judging whether life is or is not worth living amounts to answering all the fundamental questions of philosophy...I see many people die because they judge that life is not worth living. I see others paradoxically getting killed for the ideas or illusions that give them a reason for living (what is called a reason for living is also an excellent reason for dying). I therefore conclude that the meaning of life is the most urgent of questions.

In a classic of sociological method, relating the apparently individual, indeed eccentric act of suicide to social circumstances, Emile Durkheim outlined three forms – egoistic, altruistic, and (here) anomic:

There is, finally, a third sort of persons who commit suicide, contrasting both with the first variety in that their action is essentially passionate and with the second because this inspiring passion which dominates their last moment is of a wholly different nature. It is neither enthusiasm, religious, moral or political faith, nor any of the military virtues; it is anger and all the emotions customarily associated with disappointment. Brierre de Boismont, who analysed the papers left behind by 1507 suicides, found that very many expressed primarily irritation and exasperated weariness. Sometimes they contain blasphemies, violent recriminations against life in general, sometimes threats and accusations against a particular person to whom the responsibility for the suicide's unhappiness is imputed. With this group are obviously connected suicides which are preceded by a murder; a man kills himself after having killed someone else whom he accuses of having ruined his life. Never is the suicide's exasperation more obvious than when expressed not only by words but by deeds. The suicidal egoist never yields to such displays of violence. He too, doubtless, at times regrets life, but mournfully. It oppresses him, but does not irritate him by sharp conflicts. It seems empty rather than painful to him. It does not interest him, but it also does not impose positive suffering upon him. His state of depression does not even permit excitement. As for altruistic suicides, they are quite different. Almost by definition, the altruist sacrifices himself and not his fellows. We therefore encounter a third psychological form distinct from the preceding two.

This form clearly appears to be involved in the nature of anomic suicide. Unregulated emotions are adjusted neither to one another nor to the conditions they are supposed to meet; they must therefore conflict with one another most painfully. Anomy, whether progressive or regressive, by allowing requirements to exceed appropriate limits, throws open the door to disillusionment and consequently to disappointment. A man abruptly cast down below his accustomed status cannot avoid exasperation at feeling a situation escape him of which he thought himself master, and his exasperation naturally revolts against the cause, whether real or imaginary, to which he attributes his ruin. If he recognizes himself as to blame for the catastrophe, he takes it out on himself; otherwise, on someone else. In the former case there will be only suicide; in the latter, suicide may be preceded by homicide or by some other violent outburst. In both cases the feeling is the same; only its application varies. The individual always attacks himself in an access of anger, whether or not he has previously attacked another. This reversal of all his habits reduces him to a state of acute over-excitation, which necessarily tends to seek solace in acts of destruction. The object upon whom the passions thus aroused are discharged is fundamentally of secondary importance. The accident of circumstances determines their direction.

It is precisely the same whenever, far from falling below his previous status, a person is impelled in the reverse direction, constantly to surpass himself, but without rule of moderation.

Suicide: a study in sociology Paris 1897 trans. J. A. Spaulding
and G. Simpson, Routledge 1952

Local criticism and larger judgments

Extracts from notes by 1971 freshmen at East Anglia

On lines 70 ff (4 views)

'Annulled'...is a heavy, blunt sounding word receiving full stress, and is contrasted with the 'objects of delight'. Also the words 'my grief' show he is still self conscious and self pitying. Throughout the play Samson has to move to a proper state of mind which includes penitence, a recognition of the nature of his past fault and the justice of his present fate...The image of the worm is also fitting because it expresses the thoughts of a blind man who thinks he is a blind animal, crawling and creeping around in the darkness – surrounded on all sides by its darkness. There is an element of the traditional Old Testament type prophet's voice: of discordant, hollow droning, where instead of a slight fall in intonation at the end of a line, it continues at a similar pitch, falling at start of the next line, then swelling again.

> And all her various objects of delight
> Annulled, which might be...
> The sun to me is dark
> And silent as the moon.

It is difficult to know how to proceed. One realises the purposes of 'light' and sucklike images, but they are so closely interwoven it is difficult to give a clear and ordered interpretation; which may be an indication of the artist's degree of success (or failure).

Again the contrast he attempts to make between 'to live a life half-dead, a living death', does not ring true. In fact it seems as though Milton is filling out the metre.

These devices [eg alliteration and assonance on *l*] give the speech an almost lyrical air and have an effect of lulling gentleness at times, in harsh contrast to the hard repetition of the word 'vilest' or the biting line 'To daily fraud, contempt, abuse and wrong'. There is a marvellous restrained power lurking behind the bitter phrases...The function of this [To live a life half dead etc.] is obscure as its previous lulling effect would here be inappropriate. Perhaps it is just that the consonant *l* has such a round reverberant sound capable of extension so that, speaking these lines, it is easier to give them adequate emphasis and tone. They also give the impression that Samson is going round and round in endless circles beating his head against his own obtuseness. He is actually voicing his most important loss, that of grace, but consciously is relating it merely to his blindness'...

On *1034ff*

The contrast between lines 1035 and 1037, from 'virgin veil' to 'thorn intestine', 'cleaving mischief', is almost offensive in its contrast of sublime to crude:

> by her charms
> *Draws* him awry enslaved
> with *dotage* and his sense *depraved.*

...I felt Samson seemed to develop his ideas of God's favour returning almost as an excuse to find death. There is conflict between Samson's ideas and those of the other characters but no logically developing thread.

On *1574ff*

The poem is plaited with three threads. These could be subtitled 'body', 'pain' and 'peace'...The 'body' theme...generally shows Samson's power to be mortal; however strong he is, he is not strong *enough*, because by the very nature of the strength, it is necessarily limited. Samson feels bound by this inevitability, his mortalness. His strength rears against itself and so does his penis, to his destruction...

> Abortive as the first-born bloom of spring
> Nipped with the lagging rear of winter's frost

...'Abortive' as the first-born holds back in a mangled way, held up on the soft barrier made by *bs*. The *b* works in the same way as in 'orb', 'bond', 'body', 'obvious'. The fullness of the flesh here is not the cool fullness of snow. The body is thickly twisted and restricted by inevitable deformity as it begins with 'abortive'...looks as though it is about to show some soft thing, even though exposed and fallible, but it does not even get that far and ends in acute pain...The poem seems to dip in and out of revelation into complete bafflement...The poem is blocked up by a heavy progression of cancellations and there is not room to move...'Peace and consolation' are briefly revealed in the cloudy muddle of the last chorus. These don't come out fully as the resolution. It fizzles out with 'hath dismissed' condensed into *hissed*; 'And calm of mind' is vulnerably soft. 'Mind' in this poem has so far always been associated with 'hurt'...

Resources

There is a lot of material on *SA*. This list is restricted to what is comely and reviving, not of the mill. See also works cited in appendix.

Editions

BULLOUGH, G. and M. ed. *M's dramatic poems* 1958.
MACCAFFREY, ISABEL G. ed. *SA and the shorter poems of M* 1966.

Criticism

BROADBENT, J. B. *M: Comus and Samson* 1960 (Studies in literature)

BROADBENT, J. B. and LORNA SAGE *Milton* 1971 (Sussex Tapes A3 recorded discussion of *Comus, PL, SA*, with study notes and bibliography). Rept in *English poetry* ed. Sinfield 1976.

BURKE, KENNETH 'The imagery of killing' *Hudson rev* I 1948 repr as 'The "use" of M's Samson' in his *A grammar of motives* 1950.

CAREY, JOHN 'Sea, snake, flower and flame in *SA*' *Modern language rev* LXII 1967. (See McCarthy below.)

FELL, K. 'From myth to martyrdom' *English studies* (Amsterdam) XXXIV 1953.

MCCARTHY, B. EUGENE 'Metaphor and plot in *SA*' *M quarterly* VI no 4, Dec 1972. (Replies to Carey above.)

SADLER, LYNN V. 'Typographical imagery in *SA*: noon and the dragon' *English literary history* XXXVII 1970.

STEIN, ARNOLD *Heroic knowledge: an interpretation of PR and SA* Minneapolis 1957.

TAYLER, EDWARD W. 'M's firedrake' *M quarterly* VI no 3, Oct 1972.

Sources and literary analogues

Judges xiii–xvi with really large scholarly aids such as the *International critical commentary, Cruden's concordance to the Bible*, Hastings' *Dictionary of the Bible*; various psalms of dejection eg xliii; *Job* vii; *Luke* i, xi; *Hebrews* xi; prayer book for lessons etc about mid-April. See A. SMYTHE PALMER *The Samson saga and its place in comparative religion* 1913.

Compare your own attempt to create an analogue with some of these: D. H. Lawrence's story *Delilah and Mr Bircumshaw*, de Vigny's poem *La colère de Samson*, Robert Graves' poem *Angry Samson*, Eliot in *East Coker* II ('intolerable wrestle' = agonistes) and III, and Abelard and Quarles quoted in Appendix.

Music and speech

Operatic section on Samson and Jonah (equated with crucifixion and sepulchre of Christ) in *Luke passion* by Telemann (Bach's more fashionable contemporary) 1720; oratorio *Samson*, with quotations from earlier poems of M, by Handel 1754; opera by Saint-Saëns 1877. Readings directed by George Rylands with music by Raymond Leppard recorded on Argo RG 544–5. But the oratorio is not one of Handel's more interesting works; and the poem invites peformance rather than reading. It is sometimes acted but it would be more rational to devise a form to suit its peculiarities, eg the thematic importance of silence/speech; the importance of sounds to a blind hero; light/dark; possibilities of mime for blindness; the sounds of mechanical things. A chorus should be more like the vocalist backing of pop music (Betty Midler's might be a good model) than a parish choir; and varied – chanting the staccato lyrics in unison, more reflective in recitative, occasionally in naturalistic disorder (eg 115-23). The chorus of women in Eliot's *Murder in the cathedral* is not silly because they are ordinary, diurnal; so if you have a visible chorus, avoid nightdresses and the other tendencies to somnambulism that beset productions. They need to be perceived as a distinct group, or a group-person, holding a specific relationship to Samson: relatives? consultant? his respectable alter ego?

Art

See art sections in other resources lists in this series; and, generally, ANDRÉ MALRAUX *La musée imaginaire* 1952 and K. CLARK *The nude* 1956 (ch 'Energy'), both lavishly illustrated. There are accessible depictions in or by the following (start with Rembrandt: he is the artist most interested in Samson and his dates nearly coincide with M's): bas-relief of S at mill, Naples, 12c; capital of S at pillar, Autun cathedral, France, 12c; typology on enamelled squares on the 'Verdun altarpiece' at Klösterneuberg, near Vienna, by Nicholas of Verdun c. 1181; Cranach fl. 1500; woodcuts by Dürer fl. 1500; Mantegna d. 1506; foxes by Lorenzo Lotto d. 1556; Titian fl. 1540; Tintoretto fl. 1560; Guido Reni and Rubens fl. 1600; Van Dyck fl. 1630; various paintings with Delilah by Rembrandt 1606-69; Rex Whistler (1905-44) drawing repr in *Ruins* by M. Felmingham and R. Graham 1972 and *Sunday Times* 16 July 1972. Lotto is worth exploring in more detail: his four Samson designs, all c. 1530, for inlaid woodwork in the choir of S. Maria Maggiore, Bergamo, are not easily accessible but worth attempting because they are part of an OT series. He also has an extraordinary emblem in the church called *Samson betrayed by Delilah*: Samson's head peers over a millstone bulging from the ground; over them hang bunches of hair, and scissors in a pubic shape.